3rd International Conference on Vibration, Sound and System Dynamics (ICVSSD 2019)

IOP Conference Series: Materials Science and Engineering
Volume 815

Batu Ferringhi, Malaysia
2 December 2019

ISBN: 978-1-7138-2802-0
ISSN: 1757-8981

Printed from e-media with permission by:

Curran Associates, Inc.
57 Morehouse Lane
Red Hook, NY 12571

Some format issues inherent in the e-media version may also appear in this print version.

Institute of Physics
Dirac House, Temple Back
Bristol BS1 6BE UK

Phone: 44 1 17 929 7481
Fax: 44 1 17 920 0979

techtracking@iop.org

Additional copies of this publication are available from:

Curran Associates, Inc.
57 Morehouse Lane
Red Hook, NY 12571 USA
Phone: 845-758-0400
Fax: 845-758-2633
Email: curran@proceedings.com
Web: www.proceedings.com

TABLE OF CONTENTS

VIBRATION

SOUND AND ACOUSTIC

SYSTEM DYNAMICS

FLUID STUCTURE INTERACTION

IOP Conf. Series: Materials Science and Engineering **815** (2020) 011001 doi:10.1088/1757-899X/815/1/011001

ICVSSD 2019

2019 3rd International Conference on Vibration, Sound and System Dynamics (ICVSSD 2019)

2nd December 2019

Park Royal Resort, Batu Ferringhi

Penang, Malaysia.

Organized by: TheVibrationLab, School of Mechanical Engineering, Universiti Sains Malaysia (USM)

Co-organized by:

Universiti Kuala Lumpur Malaysian Spanish Institute (UniKL-MSI)

Kolej Universiti Tunku Abdul Rahman (TARC)

Universiti Tenaga Nasional (UNITEN)

2019 3rd International Conference on Vibration, Sound and System Dynamics (ICVSSD 2019)

SPONSORS

Platinum Sponsor

Collaborative Research in Engineering, Science and Technology (CREST)

CREST
COLLABORATIVE RESEARCH IN ENGINEERING, SCIENCE & TECHNOLOGY
CATALYST FOR MALAYSIA'S E&E

Silver Sponsors

MSC Software (S) Pte Ltd

MSC Software

Institution of Engineering and Technology (IET)

IET The Institution of Engineering and Technology
IET Internet of Things Network
United Kingdom

Invicom Test and Measurement Sdn. Bhd.

INVICOM
TEST & MEASUREMENT

DAG Technologies Sdn. Bhd.

DAG
TECHNOLOGIES

IOP Conf. Series: Materials Science and Engineering **815** (2020) 011001 doi:10.1088/1757-899X/815/1/011001

2019 3rd International Conference on Vibration, Sound and System Dynamics (ICVSSD 2019)

INTRODUCTION

TheVibrationLab, USM is actively pursuing research in the fields of Vibration, Sound and System Dynamics Engineering. The laboratory in collaboration with UniKL-MSI, TARC and UNITEN proposes to conduct the 3rd ICVSSD conference for the year 2019 on 2nd December 2019 located at the Park Royal Resort, Penang.

The conference is intended to bring all the researchers from universities and industries to share their knowledge in the area of research and development. It is anticipated that the conference will also be beneficial for the researchers and graduate students to improve their quality of research work. The theme of this conference for year 2019 is **" IoT and Big Data Analytics in Vibration and System Dynamics"** to encourage presenters to highlight their research work that can lead to improved competitiveness and well-being of the industry and society.

AIM

To bring together researchers in the area of **vibration, sound, system dynamics and other related areas** to share their latest findings in the area.

OBJECTIVES

- To increase the interaction and communication between the university and industry researchers and graduate students.
- To expose the researchers and graduate students in scientific and technical communication in vibration, sound and system dynamics areas
- To promote TheVibrationLab USM research activities and encourage publication works

IOP Conf. Series: Materials Science and Engineering **815** (2020) 011001 doi:10.1088/1757-899X/815/1/011001

2019 3rd International Conference on Vibration, Sound and System Dynamics (ICVSSD 2019)

CHAIRMAN

Professor Dr. Zaidi Mohd Ripin

ICVSSD 2019 ORGANIZING COMMITTEE

Vice Chairman: Dr. Ooi Lu Ean

Secretary: Mrs. Norashikin Ismail

Publication: Dr. Ahmad Zhafran Ahmad Mazlan

Registration: Mrs. Azni Shahida Khalid

Technical: Mr. Wan Mohd Amri Wan Mamat Ali

Publicity: Dr. Abdullah Aziz Saad

Treasurer: Mr. Mohamad Izudin Alisah

Webmaster: Mr. Hishammudin Endan

Workshop: Dr. Nurul Farhana Mohd Yusof

ICVSSD 2019 TECHNICAL COMMITTEE

Prof. Dr. D.C. Barton (University of Leeds, UK)

Prof. Dr. Roslan Abdul Rahman (Universiti Teknologi Malaysia)

Prof. Dr. Musa Mailah (Universiti Teknologi Malaysia)

Prof. Datuk Ir. Dr. Mohd Jailani Mohd Nor (Universiti Teknikal Malaysia Melaka)

Prof. Ir. Dr. Jawaid Iqbal Inayat Hussain (Universiti Tenaga Nasional)

Prof. Ir. Ts. Dr. Asan Ghani Abdul Muthalif (Qatar University, Qatar)

Assoc. Prof. Ir. Dr. Nawal Aswan Abdul Jalil (Universiti Putra Malaysia)

Ir. Dr. Feizal Yusof (Universiti Sains Malaysia)

Dr. Chuah Hun Guan (KDU University College)

Dr. Mohamad Ezral Baharudin (Universiti Malaysia Perlis)

Dr. Ko Ying Hao (TAR University College)

Dr. Norilmi Amilia Ismail (Universiti Sains Malaysia)

Dr. Muhammad Najib Abdul Hamid (Universiti Kuala Lumpur)

Dr. Tan Wei Hong (Universiti Malaysia Perlis)

Dr. Teoh Choe Yung (TAR University College)

Dr. Zahrul Fuadi (Syiah Kuala University, Indonesia)

Dr. Ahmad Zhafran Ahmad Mazlan (Universiti Sains Malaysia)

IOP Conf. Series: Materials Science and Engineering **815** (2020) 011001 doi:10.1088/1757-899X/815/1/011001

2019 3rd International Conference on Vibration, Sound and System Dynamics (ICVSSD 2019)

LIST OF PAPERS

Paper Number	Topics	Titles
001	Vibration	Analysis of Contact Stress and Vibration of Rolling Element Bearing
002		Preliminary Evaluation of Active Tremor Cancellation Spoon for Patients with Hand Tremor
003		Structural and Random Vibration Analysis of LEDs Conductive Polymer Interconnections
004		Sound and vibration damping properties of water retted Kenaf bast fiber (WKF) and Poly (butylene succinate) PBS
005		Monitoring of gear surface roughness in angle grinder using vibration analysis
006		Frequency domain technique for characterizing spur gears defect pattern
007	Sound and Acoustics	Numerical Analysis on Vibro-acoustic Characterisation of Hybrid Honeycomb Structure with Flow Permeability
008		Transmission Loss Analysis of Simple Expansion Tube with Micro – perforated Cylindrical Panel
009		The Evaluation of Deviation in Sound Absorption Coefficient for Micro – perforated Panel
010		Oil Palm Wastes as Sustainable Sound Absorbing Particleboard
011		Sound Absorption Characteristics of Sandwich Panel Made from Double Leaf Micro-perforated Panel And Natural Fiber
012		A Qualitative Study of the Wind Noise Cavity Inside of a Two Way Radio
013	System Dynamics	Attention of Motorcycle Handle Vibration using Suspended Handlebar
014		Development of an Ultrasonic Scalpel
015		Modelling of the Human Body Under the Exposure to Recumbent Whole-body Vibration
016		Investigation of Fatigue Failure Performance of Ankle Implant Mobile Bearing using Finite Element Method
017		Effect of Femur Geometry on Hip Protectors Testing System
018		Modelling and Closed–loop Control System for Magnetorheological Elastomer Linear Actuator
019	Fluid Structure Interaction	Numerical Study of the Flow-Induced Vibration During Tunnel Surging at the Intake Section of Chenderoh Dam
020		Vibration Analysis of Chenderoh Dam Physical Model Spillway Structure due to the Effect of Water Spilling: Numerical and Experimental Validation
021		Fluid/structure Interaction Numerical Study on the Mechanical Integrity of Water Dam Reservoir Banks

IOP Conf. Series: Materials Science and Engineering **815** (2020) 011001 doi:10.1088/1757-899X/815/1/011001

022		Coastal Sediment Transport Study along Shoreline by using Smooth Particle Hydrodynamics
023		Smooth Particle Hydrodynamics Study of Sediment Profile Erosion

CHAIRMAN WELCOMING NOTES

Dear Colleagues,

The 2019 3rd International Conference on Vibration, Sound and System Dynamics (ICVSSD 2019) seeks to bring together researchers in the area of vibration, sound and system dynamics and to share their latest findings in the area. The theme of this conference is **"IoT and Big Data Analytics in Vibration and System Dynamics"** where we encourage presenters to highlight their research work that can lead to improved competitiveness and also well-being of the industry and society. The area of vibration and system dynamics including the machine and structural health monitoring is now undergoing a major technological revamp in term of its approaches to handle the data obtained from the measurement devices. With the advent of the distributed sensors (or Internet of Things, IoT) and the availability of processing power and software to manage huge data (Big Data Analytics), the chances of finding a better solution to asset management is getting better. The improvement of machine life can come from better science of which is the foundation of this conference. The development of vibration modelling will allow better understanding and prediction of the behavior of the system and also the effect on human beings. This conference will try to gather the research work in vibration, sound and system dynamics and also the development of new sensors and systems.

The application of the big data analytics is an important area and the 3rd International Conference on Vibration, Sound and System Dynamics (ICVSSD 2019) has taken the initiative to conduct a workshop on the Big Data Analytics in Condition Monitoring in order to better disseminate the idea to the industry and also to form special interest group so that the technology can be further develop within the local industry scenario.

I would like to thank you for your contribution, whether as organizing committee, technical committee, sponsors or presenters and in particular those organizations that is willing to become the co-organizer of this event.

Thank you and "Terima kasih"

IOP Conf. Series: Materials Science and Engineering **815** (2020) 011001 doi:10.1088/1757-899X/815/1/011001

2019 3rd International Conference on Vibration, Sound and System Dynamics (ICVSSD 2019)

Professor Dr. Zaidi Mohd Ripin
Chairman ICVSSD 2019
TheVibrationLab, School of Mechanical Engineering, Universiti Sains Malaysia

PEER REVIEW STATEMENT

Peer review statement All papers published in this volume of IOP Conference Series: Materials Science and Engineering have been peer reviewed through processes administered by the proceedings Editors. Reviews were conducted by expert referees to the professional and scientific standards expected of a proceedings journal published by IOP Publishing.

IOP Conf. Series: Materials Science and Engineering **815** (2020) 011002 doi:10.1088/1757-899X/815/1/011002

Peer review statement

All papers published in this volume of *IOP Conference Series: Materials Science and Engineering* have been peer reviewed through processes administered by the proceedings Editors. Reviews were conducted by expert referees to the professional and scientific standards expected of a proceedings journal published by IOP Publishing.

Analysis of Contact Stress and Vibration of Rolling Element Bearing

Amirul Akil Adman[1] and Nurul Farhana Mohd Yusof[1]*

[1]School of Mechanical Engineering, Universiti Sains Malaysia, 14300 Nibong Tebal, Pulau Pinang, Malaysia

*Corresponding author: mefarhana@usm.my

Abstract. The initial surface and operating condition of rolling element bearing are very important as this may reduce or extend the lifetime of the bearings. Bearing condition monitoring is one of the most used method to evaluate the bearing performance. In this study, the vibration level and surface deformation produced by different contact stress condition is investigated. The experiment is carried out using rolling contact rig and the roller samples are fabricated to produce line contact similar to the roller bearing. The rollers are made from four different materials to produce different contact characteristics. The finding shows that the contact stresses affect the vibration level and the surface deformation of the rolling contact.

1. Introduction

In various rotating mechanical or work machines, rolling bearings plays important role as parts of the machine. Rolling bearings is used to reduce friction produced by energy and power which enables the force acting between the moving and the fixed parts inside the mechanical process. Analysis of vibration is important in failure diagnosis and monitoring the rotating machines. During rolling operations, it is affected by friction, lubrication, material properties, and contact mechanics. Surface failure in rolling contact occur due to the critical mechanisms which are the wear and plastic deformation.

Halme and Andersson[1] diagnosed rolling contact wear and rolling contact fatigue by combining measured and interpreted data with theory. Plastic deformation, wear and fatigue cause surface degradation. Vibration level increase due to surface degradation which the wear particles formation and rolling surface roughening occur in bearing kinematic. Vrček et al.[2] described a methodology to investigate micro-pitting damage on bearing steels using a twin-disc machine to better represent mechanical components. Smooth-rough combination of specimens fasten the micro-pitting wear. Wear always occurred on the smooth surface due to stress by the rougher surface.

Yusof and Ripin[3] evaluated the surface degradation by surface roughness and waviness parameter. The results show that the surface degradation change most with the areal surface roughness parameter (S_a) and surface waviness parameter (W_a). Wang et al.[4] studied about the influence of friction and wear of the bearing on vibration under different rubber materials of the bearing lining and contact conditions. The results show that bearing vibration is affected by both friction and wear which are cause by the speed, load and temperature. The higher speed produces low friction coefficient hence results in low vibration level.

Sulka et al.[5] studied the vibration on a rotary shaft with a ball bearing roller and compared the results of damaged and undamaged rolling ball bearing by means Short-Time Fourier Transformation (STFT) and Fast Fourier Transformation (FFT). The results show that the deformation degree increases lead to increase of vibration level. Khadersab and Shivakumar[6] proposed a technique to identify the bearing faults by vibration analysis for the rotating machinery. FFT, Inverse Fast Fourier Transformation (IFFT), and spectrogram is used for analyzed bearing faults to determine the failure of machine. Kulkarni and Wadkar[7] investigated the distributed defects of the ball bearing on the vibration response. The level of vibration varies with the increase of surface deformation. When the speed increase, the vibration level reduced. As the load increases, the surface deformation and the vibration level increase. Machado et al.[8] investigated the wear of journal bearing on rotating system in time domain.

2. Methodology

2.1. Experimental setup

In this study, a test rig for measuring rolling contact surface deformation is used as shown in Figure 1. The rig is consisting of roller specimens, stepper motor, linear guide, and four ball bearings. The test rig is put directly under Alicona Infinite Focus Microscope (IFM) therefore the surface scanning can be carried out directly. Load is applied to top roller. During surface characterization, the load is removed and there is an access hole to allow the microscope to scan the surface of the specimen. Both roller specimen is supported by two ball-bearings, but the top roller is free to move in the vertical direction, and the movement in the horizontal direction is limited by linear guides and the holders. The rolling stepper motor drives the lower roller which subsequently drives the upper roller via the line contact.

The rolling stepper motor is controlled by using Arduino which is Adafruit Motor Shield v2. The speed for the stepper motor is set at 200 rpm. The delay time after a complete cycle is set at 2 second. The programming is uploaded using Arduino Software. The slow speed is necessary in order to avoid slip and dynamic loading. Higher speed will cause higher deformation rate and as most materials are strain rate dependent, this may affect the deformation of the asperities. The high speed and dynamic loading condition may affect the ball bearings. During the experiment, the contact is dry, and the nominal contact length is 5 mm. The applied load is 50 N.

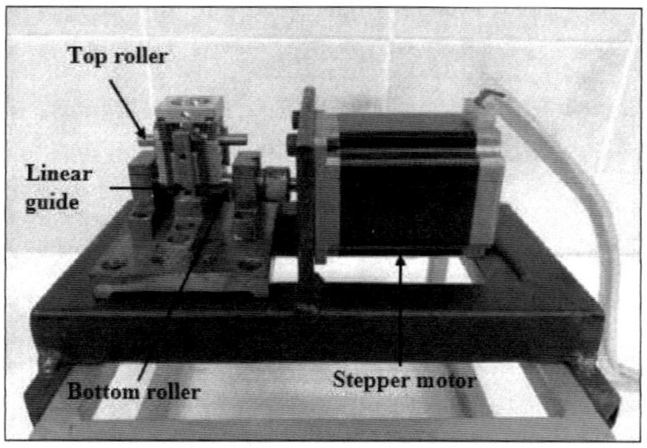

Figure 1. Experiment setup

2.2. Fabrication of roller specimens

The rollers are made from four different materials to produce different contact characteristics. Aluminium rod, brass rod, mild steel rod and stainless-steel rod were used to fabricate top roller and stainless-steel rod were used to fabricate four bottom rollers. Stainless steel was used as bottom roller to produce different contact condition with different upper roller materials. The diameter of both upper and bottom roller is 20 mm. For the fabrication of specimen, few steps have been taken in order to fit the specimens into the experimental rig. The specimen's dimension must be considered to make sure it fit into the experimental rig.

First, the metal rod of aluminium, brass, mild steel and stainless steel were cut into smaller rod using metal cutting hacksaw machine to reduce material waste. The specimen was tightly gripped into a chuck of the CNC lathe machine and the coolant was supply to the tip of the tool to avoid

IOP Conf. Series: Materials Science and Engineering **815** (2020) 012001 doi:10.1088/1757-899X/815/1/012001

damage to the surface of the specimen and also the tool. Then, the dimension of the specimen needs to be set up into the programming of the machine and then the machine was run. The machine door was closed for safety precaution which is avoid the chip from specimen leaps out.

After the bottom roller and top roller have been fabricated, the surface finish is produced. The specimen is gripped into a chuck of Conventional Lathe Machine. Then, the machine is run at 180 rev/min. The surface is being polished using 600 grit sandpaper. 600 grit sandpaper are used for all material rollers to produce the same surface roughness. The specimen is furthered cleaned with acetone solution to remove dirt.

2.3. Contact stress calculation
The contact stress that occur during the contact between two rollers made of different material is determined using Hertzian contact pressure equation [9] as shown below:

$$\text{Effective Young's Modulus,} \quad \frac{1}{E^*} = \left(\frac{1-v_1{}^2}{E_1} + \frac{1-v_2{}^2}{E_2}\right) \tag{1}$$

$$\text{Effective Radius Of Curvature,} \quad \frac{1}{R} = \frac{1}{R_1} + \frac{1}{R_2} \tag{2}$$

$$\text{Hertzian contact stress,} \; P_H = \left(\frac{PE^*}{\pi R}\right)^{0.5} \tag{3}$$

where E is Young's Modulus, v is Poison's ratio, R is radius of curvature and P is load.

From the equation, it is shown that the contact pressure is contributed by material properties: - there are Young's modulus and Poisson's ratio, load and also specimen geometry which is the radius of the rolling elements that were in contact. In this work, the roller geometry and the load are similar for all specimens and only different materials were used. Therefore, the different contact pressure obtained in all 4 tests are contributed by the material properties.

2.4. Vibration measurement
The vibration measurement is carried out using acceleration parameter where the accelerometer is used. The accelerometer is placed on the 3 different locations on the jig as shown in Figure 2. The first accelerometer C4, is placed on the upper roller mounting. The second accelerometer C2, is placed on the lower roller mounting. The third accelerometer C3, is placed on the base of the experimental rig. The acceleration level is recorded after the motor is running stably which is after a few minutes switched on the motor.

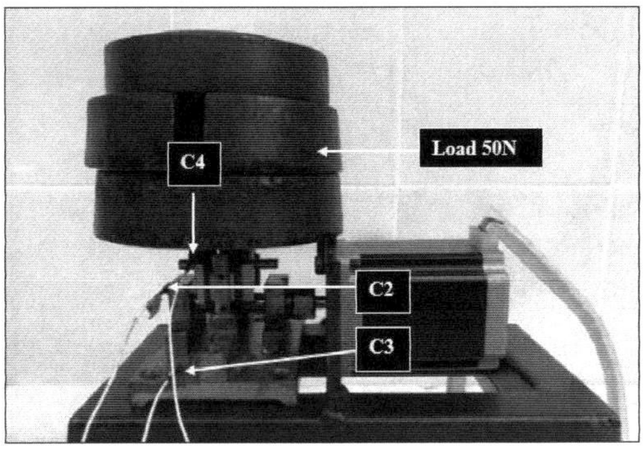

Figure 2. Location of accelerometer on the experimental jig

IOP Conf. Series: Materials Science and Engineering **815** (2020) 012001 doi:10.1088/1757-899X/815/1/012001

2.5. Surface roughness measurement

Surface scanning is carried out using Infinite Focus Microscope (IFM). The IFM is able to scan surface topography and also measure surface roughness of the surface. The roller surface roughness is evaluated using average roughness R_a parameter. It is described as the arithmetic average of the absolute values of the profile height deviations from the mean line, recorded within the evaluation length. The surface roughness of the upper roller surfaces is measured every 20 cycles up to few hundred cycles.

3. Results and discussion

3.1. Contact stress analysis

The maximum contact pressure and shear stress acting on the two cylindrical surfaces in contact is calculated using Hertzian equation. From Table 1 the configuration of surface contact between stainless steel and aluminium has the lowest contact stress compared to other materials followed by brass, mild steel and stainless steel. This shows that the Hertzian contact stress is dependent on the modulus of the elasticity of the material in contact. From calculation, it is found that as the value of modulus of the elasticity increases, the maximum contact pressure increases. Even though the stainless has the highest contact stress, it is still relatively low if compared to the yield stress. From Table 1, it is shown that aluminium has the highest contact stress/ yield stress ratio.

Table 1. Hertzian contact stress in relation to yield stress and material hardness

Material		Young's Modulus (GPa)	Hertzian Contact Stress, P_H (MPa)	Yield Stress, P_Y (MPa)	(P_H/P_Y)
Lower roller	Upper roller				
Stainless steel	Stainless steel	200	265	250	1.05
	Mild steel	200	265	250	1.05
	Brass	97	215	135	1.50
	Aluminum	69	191	20	9.55

3.2. Vibration level

Figure 3 shows the average value of vibration level at 3 different locations. The highest vibration level occur on the base of experimental rig. This is expected due to the vibration of stepper motor that is attached on the base. The vibration level at the upper roller mounting is the lowest due to the high speed of the stepper motor which result in low vibration level from the mating surfaces. The vibration level at the upper roller mounting for all materials are the same due to the vibration is produced by the contact and not by the rig. Aluminium produced highest vibration level followed by brass, mild steel and stainless steel. At high P_H/P_Y, plastic deformation occur, thus resulted in high vibration level.

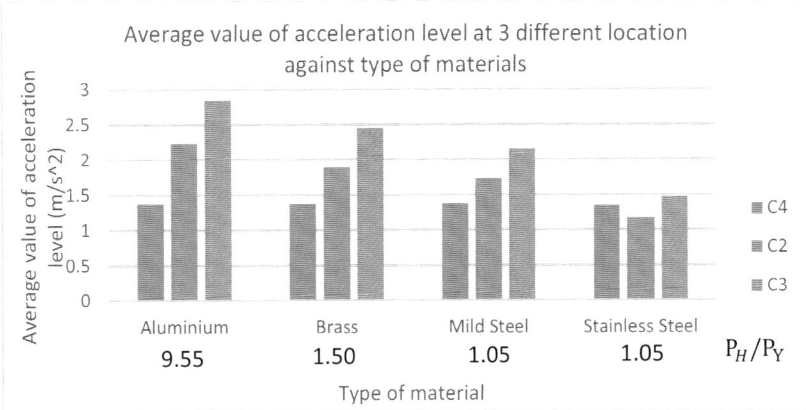

Figure 3. Average value of acceleration level at 3 different location against type of materials

3.3. Surface roughness analysis

The average surface roughness, R_a after experiment is compared with the initial R_a value as shown in Figure 4. It shows aluminium has the highest percentage of reduction followed by brass, mild steel and stainless steel. For aluminium, the high ratio between contact stress and yield stress (P_H/P_Y) cause plastic deformation drastically occur on the surface therefore reduce the R_a value for 22.8% after a few cycles. In comparison with stainless steel, the contact stress and yield stress ratio is low $\frac{P_H}{P_Y} = 1.055$ therefore the contact is partly elastic and very less plastic deformation occur. This is the reason why the R_a reduction is very small at 2.1%. It is found that as P_H/P_Y increases, the surface deformation occurred on the surfaces increases. By comparing Figure 3 and Figure 4, it is found that as the percentage of average roughness reduction higher, the vibration level also higher.

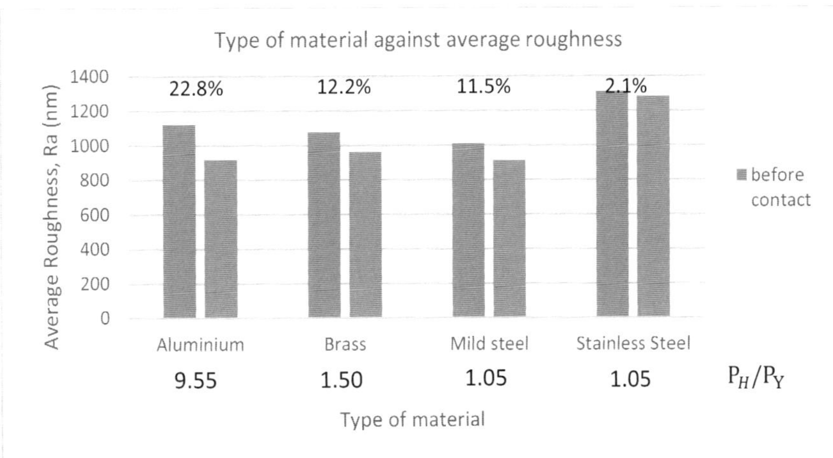

Figure 4. Average surface roughness (R_a) before and after rolling contact

4. Conclusion

The conclusions from this study are:

1. In rolling contact, the Hertzian contact stress between two mating surfaces may influence the vibration level and rate oof surface deformation. Aluminium roller with high P_H/P_Y cause high vibration level due to the contact occur at the plastic condition ($\frac{P_H}{P_Y} = 9.55$). Surface deformation also happen abruptly at this contact condition where the R_a reduce for 22.8%

after few cycles. In comparison with stainless steel roller with $\frac{P_H}{P_Y} = 1.05$, the vibration is relatively low and surface roughness reduction is small with 2.1%. This is due to less plastic deformation and the contact is partly elastic.

2. The finding shows that, the selection of materials and load is very important in machine operation. The contact condition has high influence on vibration and surface deformation.

5. References

[1] J Halme, P.A., Rolling contact fatigue and wear fundamentals for rolling bearing diagnostics – state of the art. *J. Engineering Tribology*, 2009. 224.
[2] Aleks Vrčeka, T.H., Yannick Baubet, Pär Marklund, Roland Larsson, Micro-pitting Damage of Bearing Steel Surfaces under Mixed Lubrication Conditions: Effects of Roughness, Hardness and ZDDP Additive. *Tribology International* 138, 2019. 239–249.
[3] Nurul Farhana Mohd Yusof, Z.M.R., Analysis of Surface Parameters and Vibration of Roller Bearing. *Tribology Transactions* 57, 2014. 715-729.
[4] Huanjie Wang, Z.L., Li Zou, Jun Yang, Influence of both friction and wear on the vibration of marine water lubricated rubber bearing. *Wear* 376-377 2017. 920–930.
[5] Peter Sulkaa, A.S., Vladimir Dekys, Ondrej Stalmach, Vibration analysis and comparison of the damaged and undamaged rolling ball bearing. *Transportation Research Procedia* 40 2019. 511–518.
[6] Khadersab A, D.S.S., Vibration Analysis Techniques for Rotating Machinery and its effect on Bearing Faults. *Procedia Manufacturing* 20 2018. 247–252.
[7] Sham Kulkarnia, S.B.W., Experimental Investigation for Distributed Defects in Ball Bearing using Vibration Signature Analysis. *Procedia Engineering* 144 2016. 781 – 789
[8] Tiago Henrique Machado, D.S.A., Katia Lucchesi Cavalca, Investigation about Journal Bearing Wear Effect on Rotating System Dynamic Response in Time Domain *Tribology International* 138, 2018.
[9] Bhushan, B., Introduction to Tribology. 2013: *Wiley*.

Acknowledgement
The authors would like to thank Universiti Sains Malaysia in providing the research fund (304/PMEKANIK/6315295) and workspace to complete this study.

IOP Conf. Series: Materials Science and Engineering **815** (2020) 012002 doi:10.1088/1757-899X/815/1/012002

Preliminary Evaluation of Active Tremor Cancellation Spoon for Patients with Hand Tremor

Zaidi Mohd Ripin[1]*, Ping Yi Chan[1] and Izudin Alisah[1]

[1]TheVibrationLab, School of Mechanical Engineering, Universiti Sains Malaysia, Engineering Campus 14300 Nibong Tebal, Penang, Malaysia

*Corresponding author: mezaidi@usm.my

Abstract. Hand-arm tremor is common in Parkinson's disease (PD) patients and essential tremor (ET) and can result in difficulty of carrying out daily tasks such as eating and writing. One of the approaches to overcome this is the application active tremor cancellation (ATC) to attenuate the vibration due to the hand tremor on the device hold by the hand. There are several ATC spoons in the market which are meant to help the patient to eat however third-party evaluation of the effectiveness of this technology could not be found in the literature. This study evaluates Liftware, a spoon with ATC by measuring the acceleration performance of the spoon over the frequency range of 1-12 Hz which adequately covers both the PD and ET. In the first part of the study, the device was evaluated on the isolation index throughout the tested range of frequency along the vertical or gravitational, and the horizontal directions, the direction perpendicular to both gravitational and the longitudinal axis. Subsequently, the device was evaluated for the limit at which the food it contains start to spill. Solid food, typically boiled peas and liquid, namely curry was the food tested with the spoon. Ten measurements were carried out for each type of food and the average limits prior to spillage was presented. The results show that the isolation indices in vertical and horizontal directions range from 0.06 to 0.63 and 0.06 to 0.62 respectively at frequency range of 3-12Hz. The limits of the spoon in supporting the boiled peas and curry are 1.4g and 0.9g respectively. However, the preliminary study provides part of the knowledge useful for the potential users and researchers. Further study to subject the device under different conditions, especially in clinical settings is required.

1. Introduction

Parkinson's disease (PD) is a central nervous system disorder caused by degeneration of basal ganglia resulting in the decrease in dopamine secretion. Essential tremor (ET) is another neurological disorder, which is twenty times more prevalent than Parkinson's disease. The tremors of these two disorders can progress until the daily activities of the patients are affected. Essential daily tasks such as writing, dressing, eating and drinking can become difficult to be performed when the condition is severe [1]. Assistive devices that can be used to suppress or isolate tremor have been developed to ease the conditions of the patients.

One example of the assistive devices is the Neater Eater by Michaelis Engineering [2] where the device is required to be mounted on table. Tremor is isolated and absorbed by the tremor absorber in this device. This device solves the eating problem of people who have tremor or ataxia. The limitation of this device is that its usage is confined to a same place. The passive means of applying viscoelastic material in tremor dampening has been emerged to increase fine motor functions for ET patients or individuals such that the patients are able to accurately touch an icon with an area of 1 inch by 1 inch square on a LCD panel [3].

The active tremor cancellation approach can be traced back to the research work carried out to overcome tremor in microsurgery due to physiological tremor. The obvious solution is to isolate the hand from the surgical tool by using robot assisted microsurgery which can overcome the problem of the tremor of the human hand as shown in the adoption of a dual arm six-degrees-of-freedom robot for master-slave teleoperating procedure [4].

The John Hopkins University steady-hand robot [5, 6] was designed to assist the surgeon in microsurgery. The positioning precision is reported to be approximately 5 microns. The system is further developed resulting in the second prototype with the capability of the robot to perform cannulation on approximately 80 μm outer diameter vein using biological model system [7].

Several active researches have been carried out on tremor cancellation mostly related to the cancellation of tremor in handheld microsurgical instrument which is pioneered by the study of the delicate retinal procedures where the data from the MEMS sensor and rate gyro were used to calculate the tip displacement error [8]. This method was further developed to compensate the oscillation at the instrument [9].

Saxena and Patel (2011) briefly described the expectations of a system to cancel hand tremor, particularly the need to respond in less than half of the sampling period [10] developed a device to compensate physiological tremor. The approach taken in the research to compensate microtremor of the hands in microsurgey paved the way for active tremor cancellation devices for patients with hand tremor, where in this case the tremor amplitude is larger (tens of millimeters compared to tens of microns in microtremor)

The more recent technology which managed to penetrate the market, making accessible by patients with hand tremor is a spoon that compensates tremor based on active-tremor cancellation (ATC) principle [11]. The assistive device in the form of spoon consists of sensors that measure the hand tremor of the user and microcontroller processes the data and drives the motors to produce movement of the attachment in the opposite direction of the tremor, in both the horizontal and vertical direction within the handle. Smith (2015) [12] highlighted the fact that limited evidence is available on the effectiveness of ATC spoon. To date, no published data by third-party on the frequency dependent performance, particularly the isolation capability of the active ATC spoon, are available.

This paper aims to close the information gap by establishing some baseline data of the effectiveness of the ATC spoon in the market. In this paper, the vibration acceleration isolation capability of an ATC spoon in two-dimensional linear vibration and the severity of tremor the spoon can support before spillage of food are presented.

2. Methodology

In this study the Liftware[TM], the self-stabilizing spoon that was developed by Lift Labs (Moutainview, California) to assist users with tremors to eat was evaluated. Isolation capability assessment without any load on the spoon bowl as well as evaluation on the limit the Liftware can hold without spillage were carried out.

2.1. Isolation capability assessment

The ability of the ATC spoon to isolate vibration was measured by isolation index. The isolation index is related to the transmissibility based on the following equations:

$$\text{Transmissibility} = \left|\frac{A_0}{A_i}\right| \tag{1}$$

$$\text{Isolation index} = 1 - \text{Transmissibility} \tag{2}$$

$$\text{Isolation index} = 1 - \left|\frac{A_0}{A_i}\right| \tag{3}$$

$$\text{Isolation index} = \left|\frac{A_i - A_0}{A_i}\right| \tag{4}$$

Where
A_0: Amplitude of vibration response
A_i: Amplitude of vibration input
$|A_i - A_0|$: Reduction in amplitude

The vibrational input and response were measured by using two uniaxial accelerometers (Dytran 3224A2, mass 0.2 gram) on the handle and bowl (also called spoon head) of the spoon respectively. The acquired accelerometers were connected to the National Instrument analogue input with two channels (model NI 9234), which was mounted on a data acquisition system (model NI cDAQ-9174).

Based on figure 1, the ATC spoon was mounted on a voice coil actuator (PBA System, model CVC 60-HF-20-C0.15) and driven by a signal generator (Stanford Research Systems, model DS 335) using sine swept function in the frequency range of 1-12Hz, which is adequate to cover the frequencies of tremor of PD and ET patients [13, 14]. In the experiment, the voice coil actuator acts as the shaker.

The signals were processed using fast Fourier transform (FFT) to obtain the acceleration spectrum before deriving the isolation index. The measurement was carried out in:(1) the vertical direction (refer figure 1a), where the ATC spoon was shaken along the gravitational axis and (2) the horizontal direction where the ATC spoon was shaken along the axis perpendicular to both gravitational and spoon longitudinal axis. The process from acquiring the acceleration data to the attachment of isolation index was done in National instrument LabVIEW.

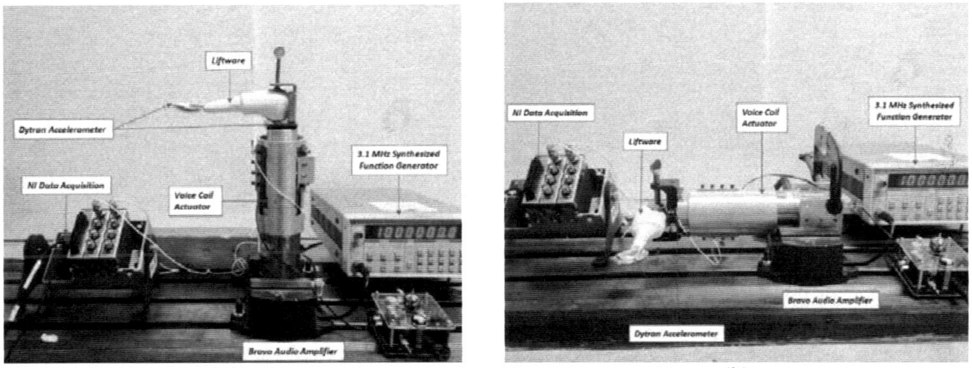

(a) (b)

Figure 1. Experimental setup of the acceleration spectra measurement for transmissibility evaluation in (a)vertical and (b) horizontal direction

2.2. Limits of ATC spoon in holding food before spillage
The peak-peak response is taken as the parameter to measure the vibration level tolerable by the spoon. Acceleration data for spillage of different types of food is not readily available in the literature, thus this experiment serves to provide such values for the determination of the severity of tremor can be tolerated when consuming different types of food. The two common types of food tested are: boiled peas and curry gravy. The similar vibrational signal consisting of a sine swept function from

1-12 Hz was applied.

The level of acceleration was gradually increased by adjusting the gain on the amplifier and the acceleration level at which the food starts to spill was marked as the maximum tolerable limit of the spoon. The spillage test on each type of food was carried out for ten times. The limits in both vertical and horizontal directions were tested separately. The mean value and the standard deviation of the ten values of maximum limit were reported.

3. Results and discussion

The frequency spectra result of the input and output vibrations of the Liftware spoon when it was subjected to vertical oscillation were shown in figure 2(a) and 2(b) respectively. By referring to figure 2(a), the input vibration peaks at around 10.7Hz, with peak amplitudes of 0.08g. At the Liftware spoon bowl, (refer to figure 2b), the peak was shifted to 10Hz, and the amplitude was reduced to 0.05g.

Figure 3 shows that isolation index varies over the spectrum. Under the tested condition, the vertical isolation index peaks at 7.9Hz and 12Hz. At these frequencies, the vibration can be isolated with Liftware for 62% and 54% respectively. At approximately 9.2Hz, the isolation index is the lowest, i.e., only 0.06. Based on the overall isolation index frequency spectrum, the Liftware can reduce 6% to 62% of tremor for the tested frequency of 3Hz or above. The performance in 3-12 Hz are more important since the PD and ET tremors [13, 14] were mainly reported in this range of frequency.

The FFT results of input and output vibrations oscillation (refer Fig. 4a and b respectively) when the Liftware was subjected to horizontal oscillation differ from those of vertical oscillation. The frequency of horizontal measurement peaks at around 10 Hz, with peak amplitude of 0.16g. At the spoon bowl, the peak occurred at around the similar frequency (figure 4b) but the amplitude was reduced to 0.11g. Figure 5 shows that the isolation index ranges from 0.06 (5.2Hz) to 0.63 (3.4Hz) for frequency above 3Hz.

The frequency spectra of the isolation index serve as design benchmark for further improvement in specific frequency range of interest, depending on the targeted tremor to isolate. The ability of the Liftware to hold 0.9g to 1.7g of vibration is reasonably usable by many patients.

When containing solid food, typically boiled peas of average 14.37gram in average, the limit of Liftware to handle vertical and horizontal vibration is approximately 17.24g and 1.42g in average of ten trials. When tested with curry gravy of 5ml, the limit of the acceleration the Liftware can handle to prevent spillage is reduced to 1.72g and 0.95g in vertical and horizontal directions in average of ten trials. The allowable limit before spillage in vertical direction is 1.2 to 1.8 times of that of horizontal direction.

Since the performance of the spoon varies in different directions, the way of gripping the spoon handle determines the superiority in isolating tremor in specific hand motion. For instance, for the case of users having tremor that is more dominant in wrist flexion-extension, it is suggested to group the Liftware by pronating the hand with palm facing down.

4. Conclusion

In conclusion, the frequency spectra of the isolation index and the limit of vibration level can be supported by Liftware before spillage provide the tangible performance of the ATC spoon. Comparison of other products and design improvement can be made based on the performance data presented.

The results were measured under the stated measurement conditions and tests on users with tremors and the measurement of the performance in three-dimensional angular parameters are required to provide a more comprehensive evaluation of the device. Nevertheless, the presented preliminary study has contributed part of the knowledge usable for the potential users and ATC device researchers and designers.

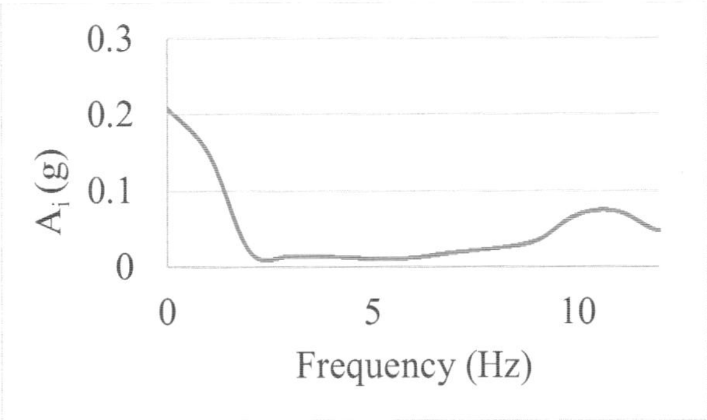

(a)

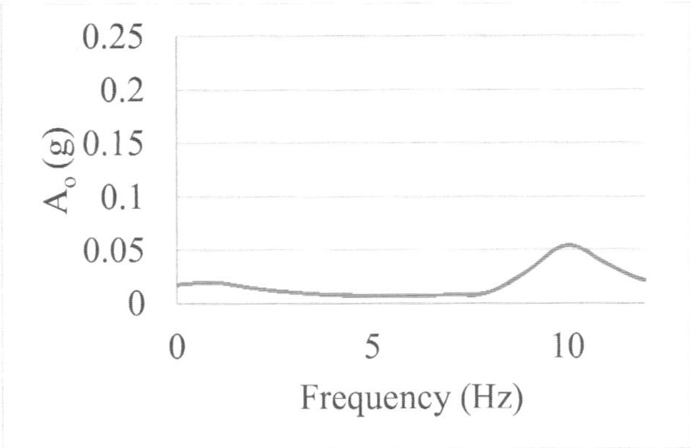

(b)

Figure 2. FFT results of (a) input and (b) output vibrations of the Liftware when it was subjected to vertical oscillation

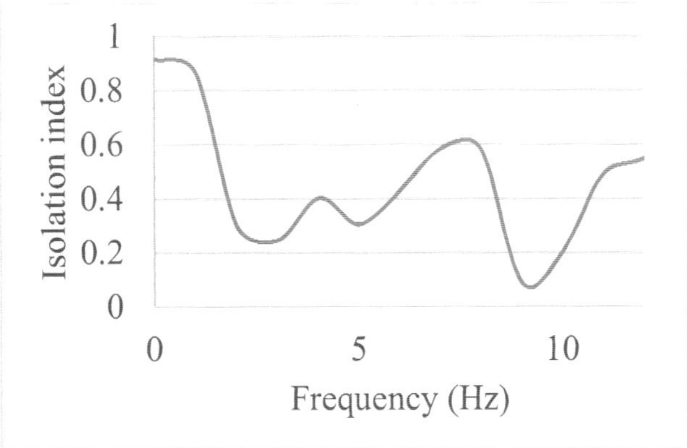

Figure 3. Frequency spectrum of isolation index of Liftware when it was subjected to vertical vibration

IOP Conf. Series: Materials Science and Engineering **815** (2020) 012002 doi:10.1088/1757-899X/815/1/012002

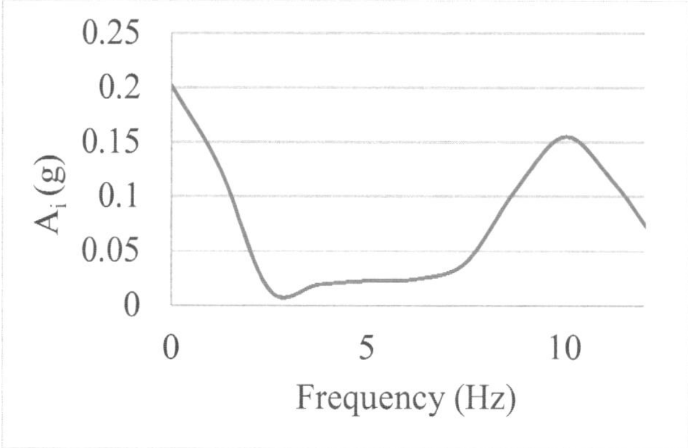

(a)

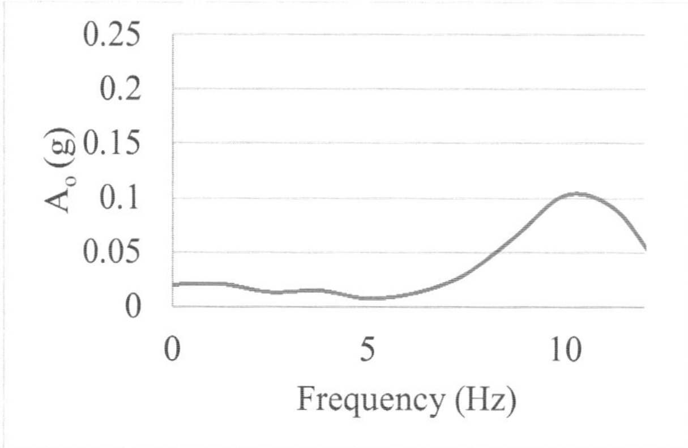

(b)

Figure 4. FFT results of (a) input and (b) output vibrations of the Liftware when it was subjected to horizontal oscillation

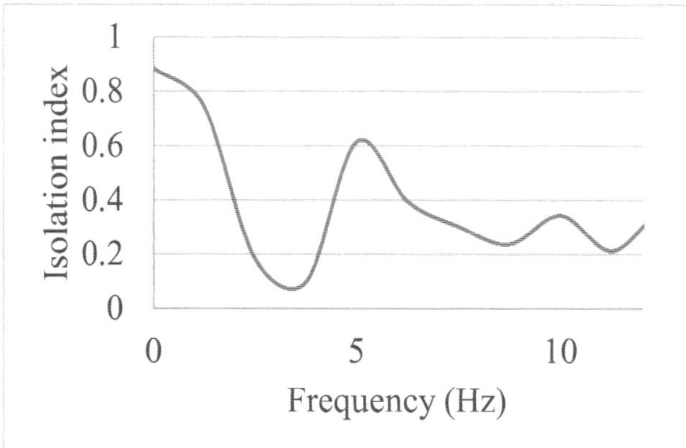

Figure 5. Frequency spectrum of isolation index of Liftware when it was subjected to horizontal vibration

Table 1. Limit of acceleration at the handle of the spoon resulted in spillage. All data were the mean or standard deviation of the 10 repetitions.

Food type	Mean amount	Vertical acceleration mean and standard deviation (m/s^2)	Horizontal acceleration mean and standard deviation (m/s^2)
Boiled peas	14.37gram	17.24, 0.32	14.17, 1.44
Soup	5ml	17.15, 3.99	9.45, 1.31

5. References

[1] L. M. Struble, "Tremors: Learning to stop the shakes," *The Nurse Practitioner: The American Journal of Primary Health Care,* vol. 35, no. 6, p. 18, 2010.

[2] J. Michaelis, "Introducing the neater eater," *Action Res,* vol. 6, no. 1, pp. 2-3, 1988.

[3] C. Brauchie, K. Hurley, G. Barrett, S. Morton, and S. Barrett, "Wrist/Arm support to assist in fine motor control for essential tremor patients," *Biomedical sciences instrumentation,* vol. 50, pp. 37-40, 2014.

[4] P. S. Schenker, H. Das, and T. R. Ohm, "A new robot for high dexterity microsurgery," in *International Conference on Computer Vision, Virtual Reality, and Robotics in Medicine*, 1995, pp. 115-122: Springer.

[5] R. Kumar, G. D. Hager, A. Barnes, P. Jensen, and R. H. Taylor, "An augmentation system for fine manipulation," in *International Conference on Medical Image Computing and Computer-Assisted Intervention*, 2000, pp. 956-965: Springer.

[6] R. Taylor *et al.*, "A steady-hand robotic system for microsurgical augmentation," *The International Journal of Robotics Research,* vol. 18, no. 12, pp. 1201-1210, 1999.

[7] B. Mitchell *et al.*, "Development and application of a new steady-hand manipulator for retinal surgery," in *Proceedings 2007 IEEE International Conference on Robotics and Automation*, 2007, pp. 623-629: IEEE.

[8] M. Gomez-Blanco, C. N. Riviere, and P. K. Khosla, *Sensing hand tremor in a vitreoretinal microsurgical instrument*. Carnegie Mellon University, The Robotics Institute, 1999.

[9] C. N. Riviere, W. T. Ang, and P. K. Khosla, "Toward active tremor canceling in handheld microsurgical instruments," *IEEE Transactions on Robotics and Automation,* vol. 19, no. 5, pp. 793-800, 2003.

[10] A. Saxena and R. V. Patel, "An Active Handheld Tremor Cancellation Device," in *ECE Grad Symp, UWO, June 2011*, 2011.

[11] C. Norris, "Liftware: Tremor Cancellation Technology. Stabilize Your Hand Tremor," p. 10, 2 December 2014. Accessed on: 11 November 2019Available: https://clairenicolenorris.files.wordpress.com/2015/03/norris_claire_white_paper_liftware.pdf

[12] N. Smith, "Liftware: Self-stabilizing Eating Utensils for Individuals with Hand Tremor," *CADTH Issues in Emerging Health Technologies*, no. 142, Oct 2015. Accessed on: 11 November 2019Available: https://www.cadth.ca/sites/default/files/pdf/EH0030_liftware_self_stabilizing_eating_utensils_for_individuals_with_hand_tremor-e.pdf

[13] P. H. G. Mansur *et al.*, "A review on techniques for tremor recording and quantification," *Crit. Rev. Biomed. Eng.,* vol. 35, no. 5, pp. 343-362, 2007.

[14] P. G. Bain, "The management of tremor," *J. Neurol. Neurosurg. Psychiatry,* vol. 72, no. suppl 1, pp. i3-i9, 2002.

Acknowledgement
This work was supported by the Ministry of Higher Education (MOHE) under Exploratory Research Grant Scheme ERGS (203/6730080) and Innovation Seed Fund USM (PMEKANIK AUPI00239).

Structural and Random Vibration Analysis of LEDs Conductive Polymer Interconnections

S Zulfiqar[1], AA Saad[1*], MA Wan Chek[1], MFM Sharif[1], Z Samsudin[2] and MYT Ali[2]

[1] School of Mechanical Engineering, Universiti Sains Malaysia, 14300 Nibong Tebal, Penang, Malaysia.

[2] Jabil Circuit Sdn Bhd, Bayan Lepas Industrial Park, 11900 Penang, Malaysia.

[*] Corresponding author: azizsaad@usm.my

Abstract. Many devices consist of electronic parts and can be operated in a certain vibration environment for some instant without undergoing failure. Interconnection or joint plays a very important role in electronic devices to connect various electrical and mechanical parts altogether. The main objective of this study is to investigate the effect of interconnection performance under vibration loading in terms of natural frequency, mode shapes and transfer function. Four different models of LEDs are built for pull strength and vibration tests. The structural analysis of these models are carried out by using universal testing machine. The dynamic responses of four build matrices under certain random frequency are observed by random vibration analysis. In structural analysis, the samples are subjected to vertical displacement loading while in random vibration analysis, the models run in a frequency range of 3 Hz to 500 Hz. It is observed that the structure of the adhesive joints affect the strength of interconnection between circuit pad and LED joints. Furthermore, FEA analysis is also carried out to determine the stresses at solder joint under vibration condition. The models are excited from natural frequency and good comparison of the experimental results and FEA analysis are observed.

1. Introduction

The electronic system such as control module, has been used by many companies, especially automation companies. These electronic components are mostly operated in severe vibration environment for certain time period without undergoing failure. An electronic system can be subjected to many different forms of vibration. The most popular form of vibration is random vibration. Random vibration is the most related vibration condition subjected to destabilization created by rotational and tumbling masses. The degree of freedom of vibration in a system with certain natural frequency is known as vibration mode. The range of frequency varies according to its application such as [1] proposed that 2.5 Ton truck has 15 G and 19 G acceleration level with 15 Hz – 40 Hz frequency range at a speed of 10 mph – 15 mph. The vibration frequency range in automobiles increases with the system, i.e., vibration while driving, suspension and steering systems. The frequency range in this case ranges from 3 Hz to 1000 Hz [2]. A range of frequency is excited at the same time which means that the frequency loading corresponds to all masses are represented at the same time and therefore, large displacement amplitudes are obtained to produce failure impact in analysis system. Power Spectral Density (PSD) input is assigned in the simulation to quantify the random frequency. PSD is the input curve of acceleration units per Hz^2 versus range of frequencies. PSD input is referred from JEDEC Standard 22b103B vibration, variable frequency. JEDEC standard is used to test the reliability of electronic devices under different vibration applications to which component can be exposed [3]. In this paper, PSD input is based on level where a component can be exposed to the most severe vibration condition. PSD excitation is applied on the fixed supports of models in y – axis direction. Furthermore, harmonic response analysis is then used to determine the response of model structure under a steady state sinusoidal (harmonic) loading at a given frequency. In harmonic analysis, the peak response corresponds with the natural frequencies of the structure to be analysed [4]. Usually, harmonic response analysis is done to obtain the values of stress equivalent, displacement and relative sensitivity of the frequency response [5].

Several electronic components are connected together by means of interconnection materials. These interconnection materials are known as solders. The most common solders are lead – based solder, lead – free solder and conductive adhesives. Solder joint failure occurs in electronic devices due to vibration loading, therefore, the structure and reliability of solder joint play a very important role. Now a days, tin – lead solder (Sn – Pb) that is lead – based solder does not longer used by electronic industries due to restriction in law and new requirements for non – toxic materials. The lead – free solders and conductive adhesives are being used to meet new environmental and green requirements [6], [7]. Lead – free solder is a tin – based alloy such as Sn – Ag, Sn – Cu, Sn – Zn, Sn – Ag – Cu, etc. Conductive adhesives are the combination of adhesive epoxy and a conductive metal such as silver, copper, gold, tin oxide, or indium. Silver is used as an epoxy filler because it is a good conductor of electricity, easily available and cheap in cost. Conductive adhesive is applied in the form of paste to an electronic circuit before placement of electronic components. It does not require particular pressure for curing and its curing temperature ranges from 130°C – 180°C which is suitable for thermal sensitive electronic components [8].

Finite element analysis (FEA) is then used to investigate the solder joint stresses and fatigue estimation. Many researchers have used two models of FEA that are detailed FE models and Smeared FE models. Smeared FE models give the detailed results of design while detailed FE models are not recommended because this model requires large amount of time period to build and solve. Such model is excessive in nature when simplified models can produced accurate data quickly with less efforts [9]. During FEA, the convergence of analysed frequency results can be examined by applying variation in mesh densities in the model. Therefore, large size meshes help to relocate the frequency results. The natural frequencies test vehicle are examined by modal testing method to verify FEA model [10]. The stress, strain, displacement, maximum or minimum load applications can also calculated by FEA simulation. FE analysis of solder joint fatigue estimation is then estimated with experimental modal analysis to determine natural frequencies and mode shapes [11]. The modal analysis is first done on the global model and then compared with the experimental results. The global local approach is applied to determine the equivalent strain in critical areas of solder joints [12].

The main objective of the present research is to study the structural and reliability of conductive polymer interconnections under vibration loading. FE analysis is also carried out in order to determine the stress generated due to vibration loading. The comparison of results obtained by experiment and simulation is also done.

2. Methodology

2.1 Pull Strength Test

Four LEDs are bonded to the circuit by using conductive adhesive (Loktite Ablestik ABP 2032S) with different build matrices. The pull strength test is used to evaluate the bonding strength of LEDs with conductive polymer interconnections. The four samples of each build matrix are tested and average readings are considered as final readings. Universal testing machine (UTM – INSTRON 3367) is used to measure the strength of interconnections between LED joints and circuit. The circuit is printed on Lexan 8010 polycarbonate sheet (PC) base and padded with silver epoxy as solder joint depends on different dispensing methods. The sheet is then cut into small pieces consisted of one LED per piece. The dispensing method for four LED models are shown in figure 1.

IOP Conf. Series: Materials Science and Engineering **815** (2020) 012003 doi:10.1088/1757-899X/815/1/012003

2.2 Vibration loading Test

All the four samples are then attached on a custom design L – shaped base aluminum sheet of 5 mm thickness as illustrated in figure 2. This platform is mounted on a shaker machine for vibration test, the frequency range of the shaker is set to 3 Hz – 500 Hz. The samples run at the same time under swept random vibration for 4 hours. Modal testing is done to determine the natural frequency of structure prior to the vibration test. The bonding between LED joints and circuit are observed at every 30 minutes time interval. The electrical conductivity before and after vibration loading of every sample is evaluated by connecting a power supply in terms of current variation. The bonding is further inspected by using Hirox Digital Microscope (KH – 3000) and x – ray machine (XT V 160) to detect any failure like cracks in the conductive paste.

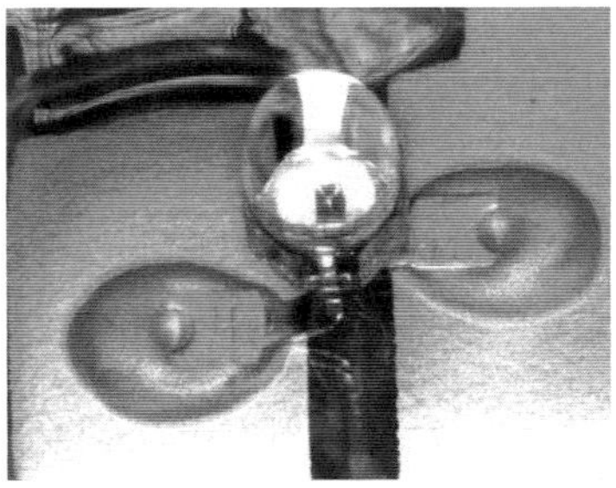

(a) Model 1. 3 dots of solder paste at the bottom + one dot of solder paste on the edge of LED lead base.

(b) Model 2. 3 dots of adhesive close in straight line on the circuit but with reduction number of dots in straight line one by one.

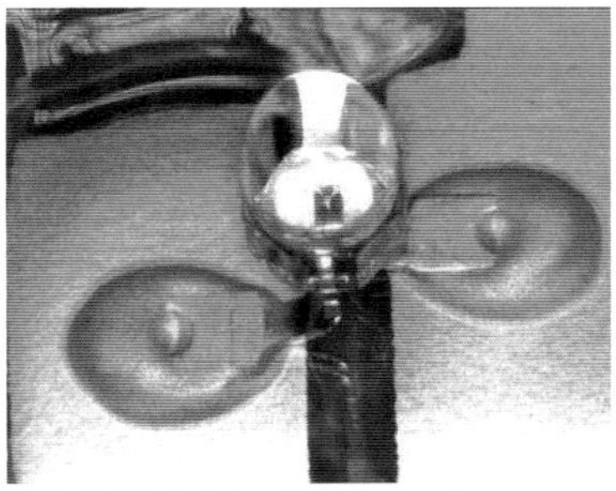

(c) Model 3. 3 dots of solder paste surrounds the sides of LED lead base

(d) Model 4. 2 dots of solder paste at each side of LED lead base

Figure 1. LED samples with solder joints

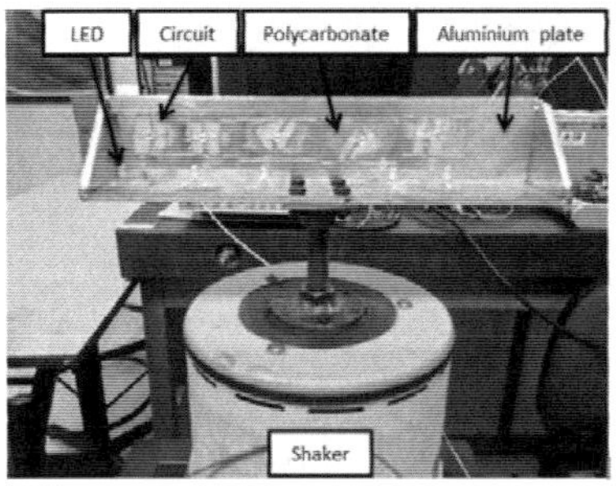

Figure 2. Vibration Test setup for LEDs joints.

IOP Conf. Series: Materials Science and Engineering **815** (2020) 012003 doi:10.1088/1757-899X/815/1/012003

2.3 Geometry of LED Models

Based on the assembly prototype model used in experimental analysis, four 3D CAD models are developed by using SolidWorks software and then imported to ANSYS 19.2 for further analysis. These CAD models are designed by considering actual component specifications from manufacturer. The actual dimensions of solder joints failure cannot be determined because solder joints do not come under geometry structure, therefore, the solder joints can approximately be established on the basis of actual shape as illustrated in figure 3. The model of copper pad circuit on PC sheet base is modified to cater the needs of simulation as discussed in table 1. However, the complete assembly model is shown in figure 4.

Table 1. Dimensions of each part of LED models

Parts	Dimensions (mm)
Base of lead	5.5 x 2.5 x 0.5
Solder joint Model 1	6.5 x 3.5 x 1.2
Solder joint Model 2	5.0 x 3.5 x 1.2
Solder joint Model 3	6.5 x 3.5 x 0.5
Solder joint Model 4	5.0 x 3.5 x 0.5

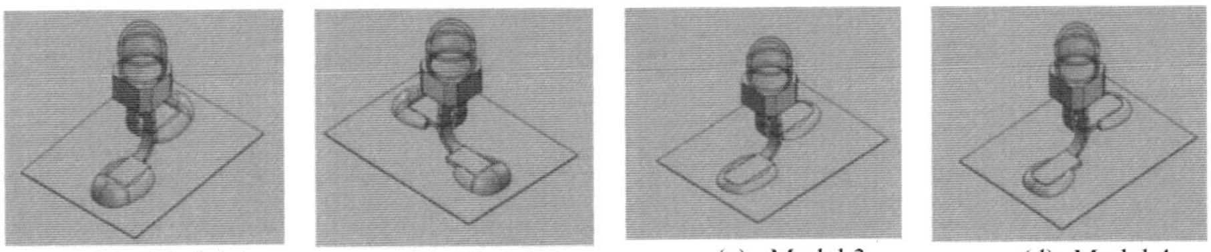

(a) Model 1 (b) Model 2 (c) Model 3 (d) Model 4

Figure 3. 3D CAD models of LED with Solder joints

The materials applied in ANSYS are based on the material used in experimental prototype model. The mechanical properties of material are provided from ANSYS materials library and manufacturer sources. The materials are assumed to be non – linear elastic – plastic in nature. The mechanical properties of materials used in different parts of LED model are summarized in table 2.

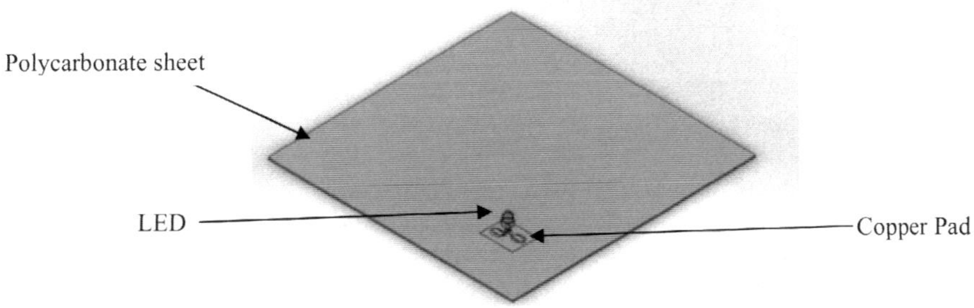

Figure 4. Complete assembly model

Table 2. Mechanical properties of materials used in different parts of LED models

Parts	Materials	Density (kg/m^2)	Elastic Modulus (MPa)	Poisson's Ratio	Yield Strength (MPa)	Tangent Modulus (MPa)	UTS (MPa)
Copper Pad circuit	Copper	8900	130000	0.34	120	125	210
Polycarbonate base	PC	1200	2506	0.38	63	.05	65
Interconnections	Silver epoxy	4500	4140	0.32	24.1	38.62	34.5
Lead	Copper alloy	8300	110000	0.34	280	1150	491

2.4. Random Vibration Analysis

Random vibration analysis is carried out in ANSYS 19.2 to analyze the failure occurred on each model of LED under random vibration loadings. Before doing this analysis, the dynamic responses on each model are identified by performing modal analysis. Only fixed support at the end surface of PC sheet base is applied as a boundary condition. In modal analysis, six different mode shapes are identified within the frequency range of 3 Hz – 500 Hz. After that, these mode shapes are used in the random vibration test for further analysis.

Random vibration is a non – deterministic motion. The vibration pattern would be varied, to quantify the frequency excitation PSD input need to be assigned in the simulation. In this analysis, band limited white noise has been used having a constant spectral density over specified frequency range as shown in figure 5.

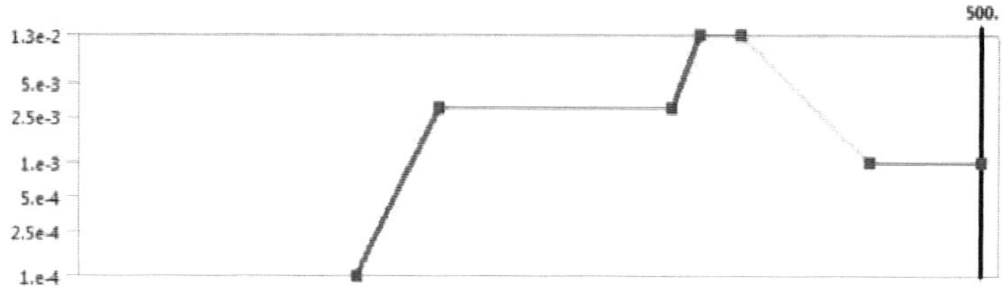

Figure 5. Power Spectral Density (PSD) input curve

PSD input is referred by JEDEC standard 22b 103B vibration that is used to test the reliability of structure under various vibration loadings to which components can be exposed [3]. In this paper, PSD input is based on the level where all components can be exposed to severe conditions. PSD excitation is applied in y – axis direction perpendicular to model plane at fixed support.

2.5. Harmonic Response Analysis

The solution from modal analysis is used as input in harmonic response analysis. The main purpose of this analysis is to determine the response of model structure under steady – state sinusoidal loading at a given frequency. The input parameters in harmonic analysis are Young's modulus, Poisson's ratio and mass density. Since modal coordinates are used, therefore, harmonic solution using the mode superposition method automatically first perform a modal analysis. The mode superposition method is much faster than other methods that's why harmonic analysis portion gives the results very quick and efficient. The peak response corresponds to natural frequency of the structure are analyzed in harmonic analysis. The number of maximum modes is used as a first boundary condition. After that, the solution interval is set at 100 intervals so that more accurate results obtained. The free vibration analysis also performed internally with mode superposition method. It is analyzed that acceleration of frequency excitation PSD input needs to be assigned in the simulation.

3.0 Results and Discussions

3.1 Pull Strength Test

It is observed that 3rd build matrix from figure 6(c) has the highest strength value of 13.16 N than 1st, 2nd and 4th build matrices that are 10.38 N, 9.69 N and 7.04 N respectively. The strength value and area of failure depend on the amount of conductive adhesive used in connecting LEDs with circuit pad. Figure 6 describes the areas of failure of four build matrices. 1st and 3rd build matrices have the largest areas as compared to other matrices because they used larger amount of adhesive. The increment of surface contact area between the adhesive and circuit lead to increase the pull strength values.

3.2 Visual Inspection of LEDs

The vibration test analysis of al LEDs build matrices is carried out. It can be observed that no disconnection occurred at conductive adhesive or circuit interface after vibration loading in both the directions, vertical and horizontal as shown in figure 7 and figure 8 respectively. Both LED joints and adhesive are not peeled off from the circuit as well as no cracks are generated in adhesive bonding.

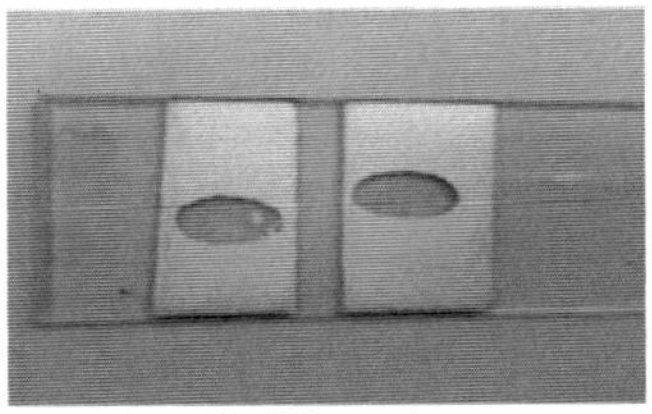

(a) 1st build matrix

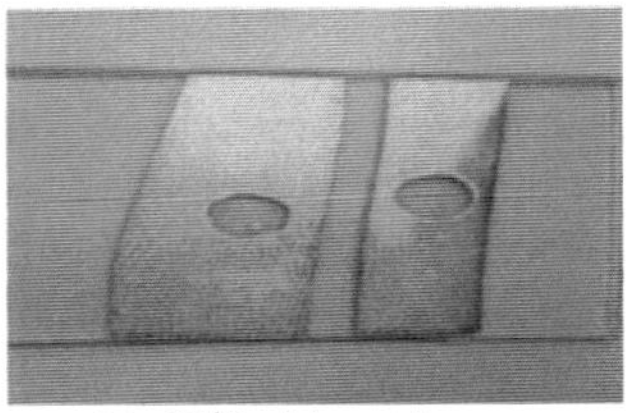

(b) 2nd build matrix

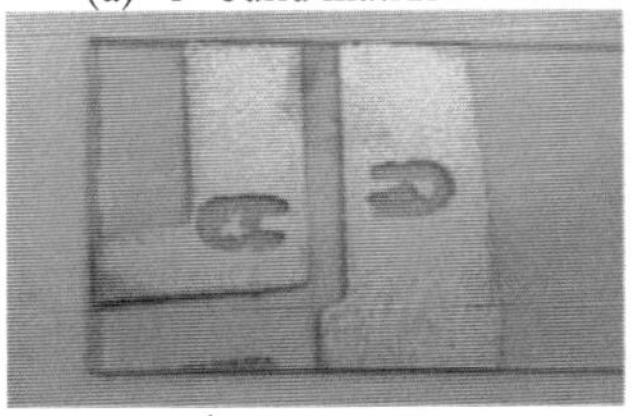

(c) 3rd build matrix

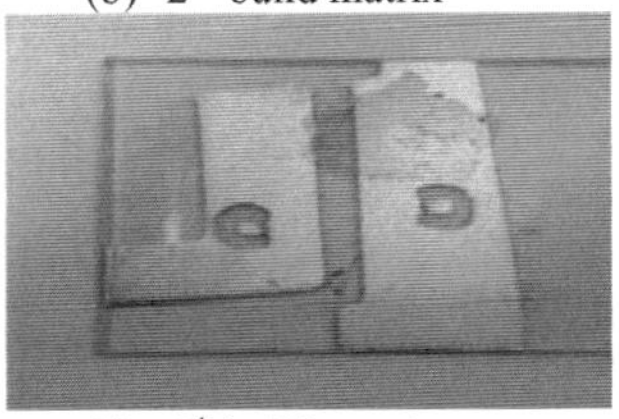

(d) 4th build matrix

Figure 6. Failed areas formation after pull test

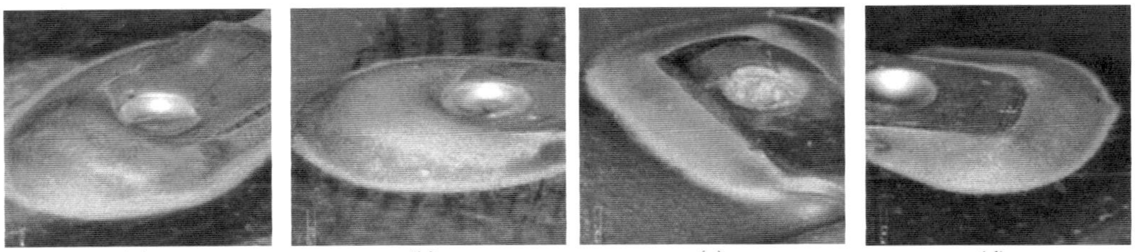

(a) (b) (c) (d)

Figure 7. Bonding at LED joints on vertical plane for (a) 1st build matrix (b) 2nd build matrix (c) 3rd build matrix (d) 4th build matrix

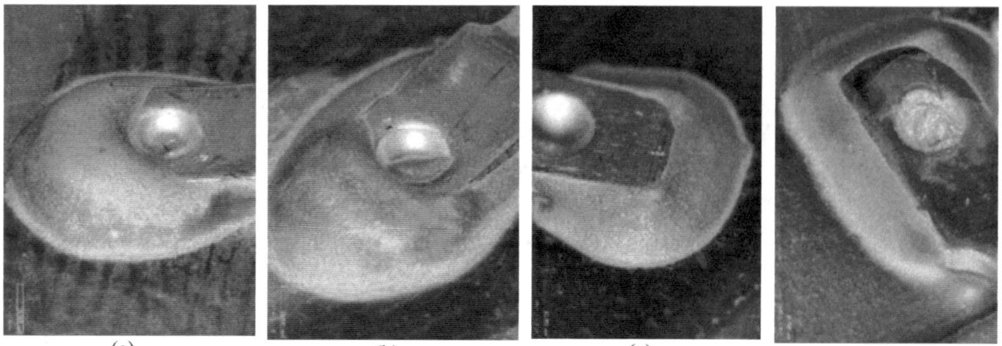

(a) (b) (c) (d)

Figure 8. Bonding at LED joints on horizontal plane for (a) 1st build matrix (b) 2nd build matrix (c) 3rd build matrix (d) 4th build matrix

3.3 Structural Analysis

The interconnection strength and structure are analysed using ANSYS 19.2. The simulations of pull strength test for four models are carried out. The maximum deformation values of each model obtained from simulation are slightly different from that of pull strength test because the dimensions of solder joint are not exactly same as in experiment. These deformation values from simulation and experimental are constantly increasing with small deviations as described in table 3. Hence, FEA model is verified and the results are reliable for further analysis.

Table 3. Maximum Deformations of 4 models

Models	Time (sec)	Maximum Deformation (mm)		Percentage Error (%)	Min. Deformation (mm)	Max. Deformation (mm)
		Simulation	Experimental			
1	4.5	0.10626	0.10832	1.94	0.07228	1.5327
	7.3	0.17066	0.16689	2.22		
	19.3	.0461025	0.44175	4.19		
	35.6	0.87866	0.83341	5.15		
2	12	0.25287	0.24796	1.15	0.25287	1.2328
	22.8	0.46574	0.44646	4.13		
	30.9	0.62863	0.58210	7.40		
	43.1	0.87776	0.83560	4.80		
3	5.3	0.15421	0.15293	0.47	0.01039	2.0551
	12.4	0.38495	0.33946	11.69		
	18.8	0.59982	0.60166	0.31		
	26.8	0.87490	0.89646	2.46		
4	6.0	0.13372	0.13350	0.16	0.13372	1.2582
	12.0	0.25357	0.23304	8.10		
	21.0	0.43438	0.41636	4.15		
	34.5	0.71295	0.72525	1.73		

Model 3 has the highest maximum deformation than others because different designs of adhesive joints affect the overall equivalent stress and total deformation in a pull test simulation. Figure 9 shows FE analysis of total deformation for each model.

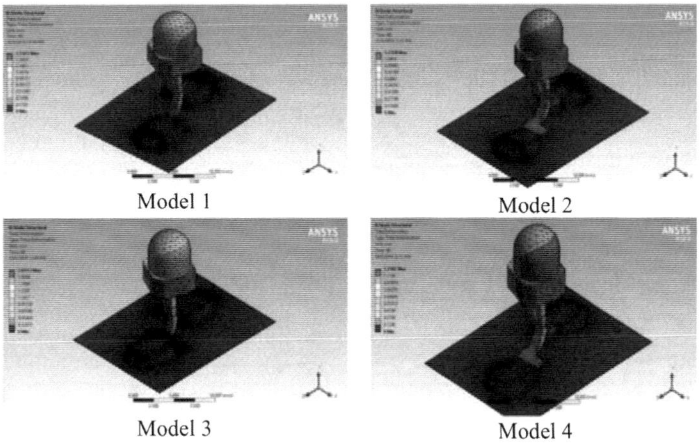

Figure 9. Total deformations of LEDs models

Similarly, the stress analysis is done for each model shown in figure 10. The maximum equivalent stresses of all models mostly concentrated at the corner of base LED. The maximum equivalent stresses of four models (Model 1 to 4) are 489.16, 363.56, 632.68 and 360.92 MPa repectively. LED model 3 shows the highest equivalent stress value than others and is most likely to fail at the corner of LED lead because equivalent stress exceeds UTS (Universal tensile stress) values. Furthermore, the stress analysis of the interconnections of four LED samples is also conducted. Figure 11 shows the maximum stress occurred at the edges of solder joints. The values of maximum equivalent stresses of adhesive joints are 51.394, 36.967, 51.183 and 38.690 MPa of four Models.

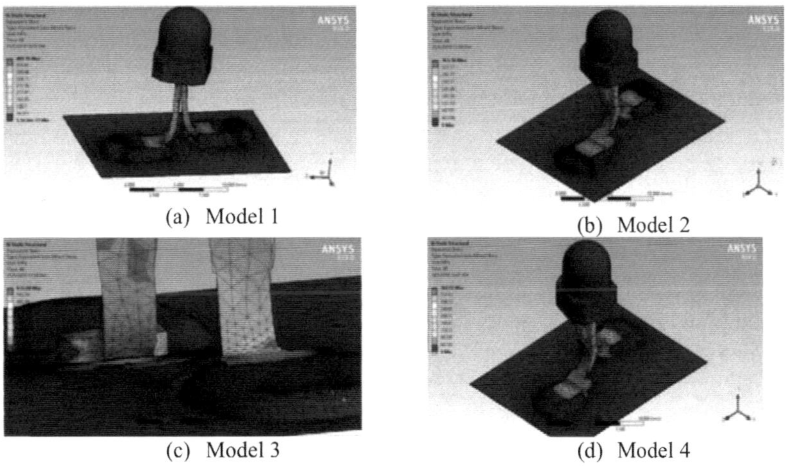

Figure 10. Maximum Equivalent stresses of full body

From figure 11, it is observed that models 1 and 3 have the highest values of equivalent stress than that of models 2 and 4 which means that adhesive joint structure of models 2 and 4 are stronger than that of models 1 and 3. The maximum stress values of adhesive joints materials of each sample are compared to UTS values and is concluded that all adhesive joints fail but the most suitable models are 2 and 4 that have smaller maximum stress values. According to experimental results, some LED assembly have voids at certain places and adhesive solder joints make fracture and failure.

IOP Conf. Series: Materials Science and Engineering **815** (2020) 012003 doi:10.1088/1757-899X/815/1/012003

The pull strength test simulation also emphasizes on the stress concentration of copper pad circuit. If the adhesive joint and LED has higher maximum stress equivalent then copper pads will be effected by higher stress values. The obtained values of maximum stress of copper pad circuit of models 1-4 are 105.57, 143.64, 142.51 and 146.53 MPa respectively. The simulation of copper pad is shown in Figure 12.

As a result, model 4 has the highest maximum equivalent stress value than other models, In order to identify whether the failure occurred or not, the maximum stress value of copper material is compared with its UTS value. Despite of all, the LED models are suitable to utilize the copper pad circuit. Experimentally, a model can fail at solder joint or copper pad circuit due to large difference in material properties and thickness of copper pad circuit used in simulation.

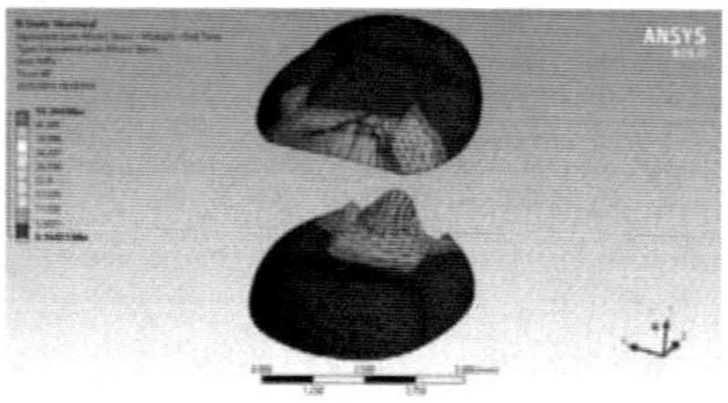

(a) Model 1

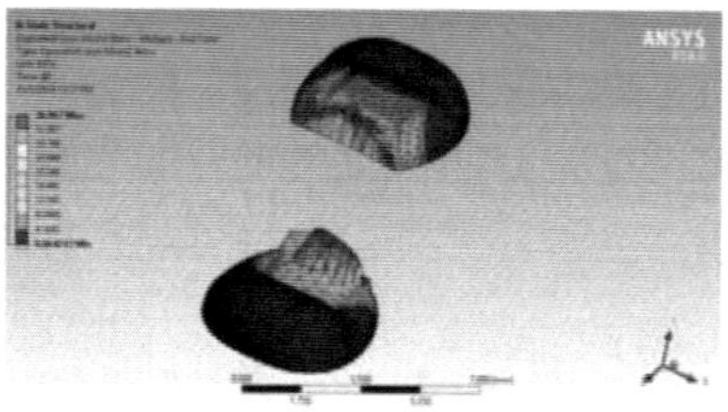

(b) Model 2

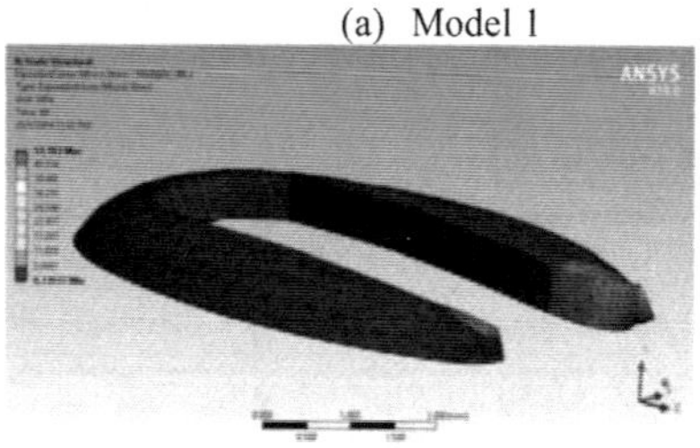

(c) Model 3

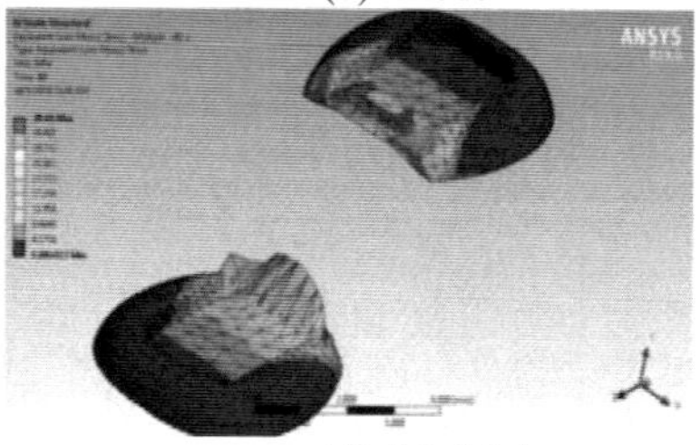

(d) Model 4

Figure 11. Maximum Equivalent stresses at solder joints

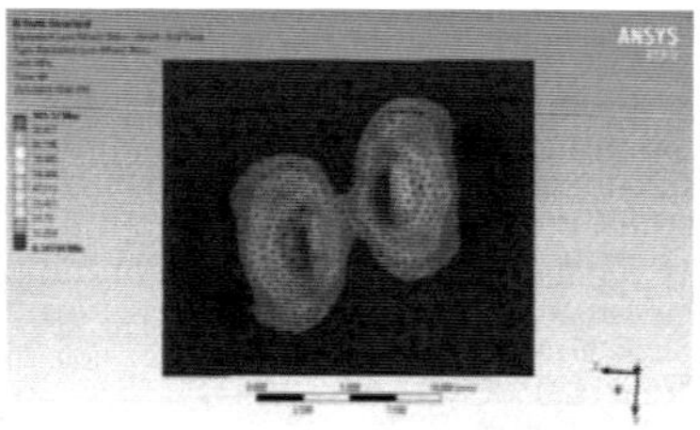

(a) Model 1

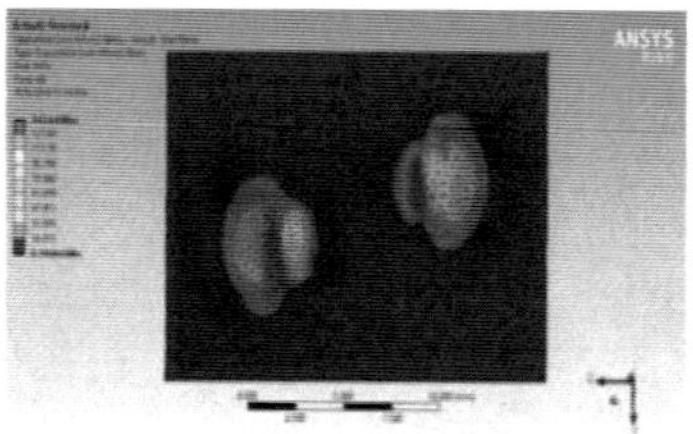

(b) Model 2

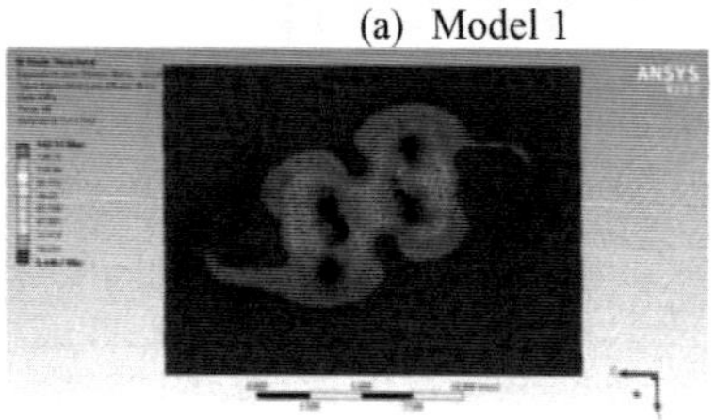

(c) Model 3

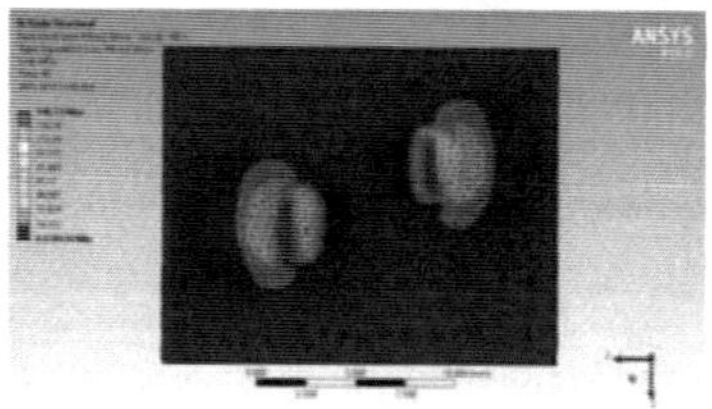

(d) Model 4

Figure 12. Maximum Equivalent stresses at copper pad circuit

IOP Conf. Series: Materials Science and Engineering **815** (2020) 012003 doi:10.1088/1757-899X/815/1/012003

3.4 Modal Analysis

Modal analysis is carried out in order to determine the dynamic response of every model at specific frequency and range of deformation occurs on each model within the frequency range. 6 modes are generated in this analysis. The natural frequencies of these modes computed by ANSYS, are in the range of 3 Hz – 500 Hz. These modes are utilized to create the mode shape results. The deformations for each and every selected frequency are then calculated. All models have the same pattern of mode shapes and number of frequencies. 6 mode shapes with respect to frequency range 3 Hz – 500 Hz are shown in figure 13.

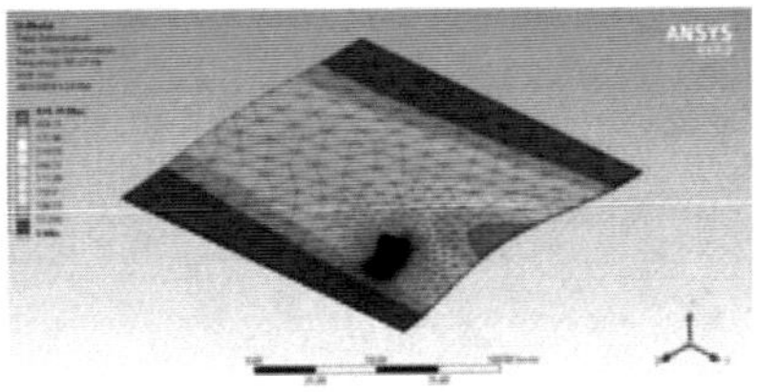

(a) Mode shape 1 ($f = 97.138\ Hz$)

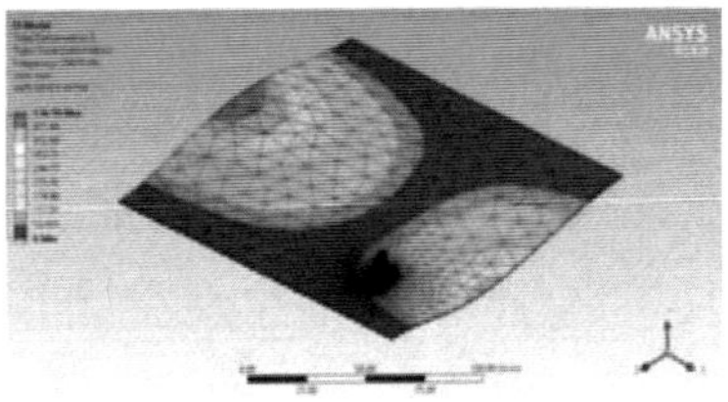

(b) Mode shape 1 ($f = 126.67\ Hz$)

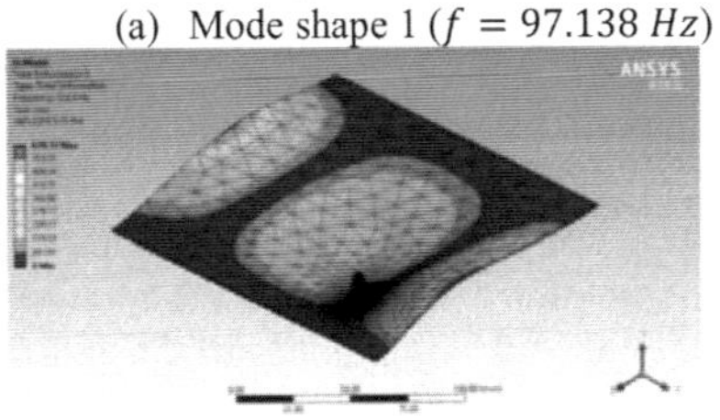

(c) Mode shape 1 ($f = 212.33\ Hz$)

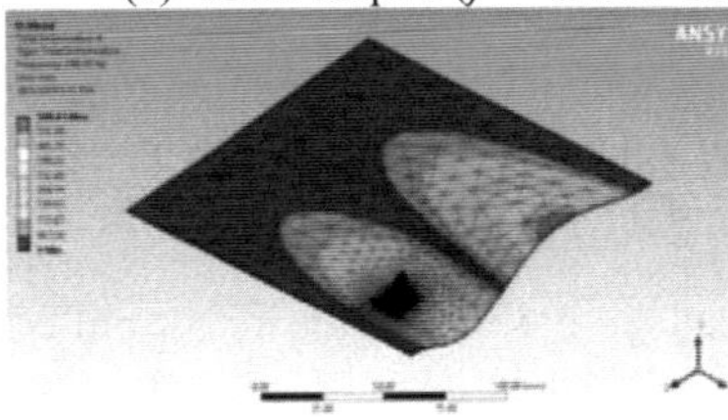

(d) Mode shape 1 ($f = 288.21\ Hz$)

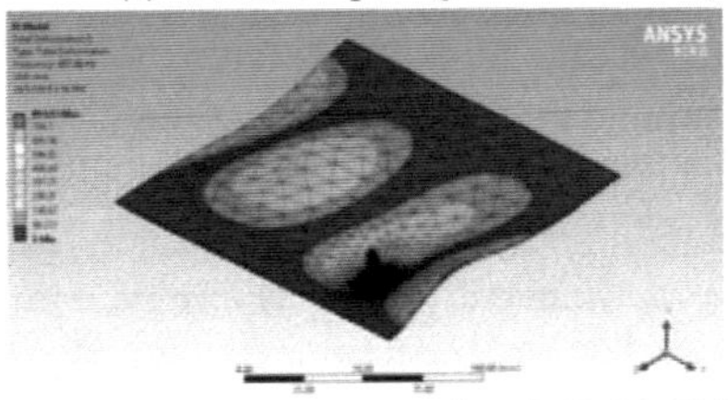

(e) Mode shape 1 ($f = 368.11\ Hz$)

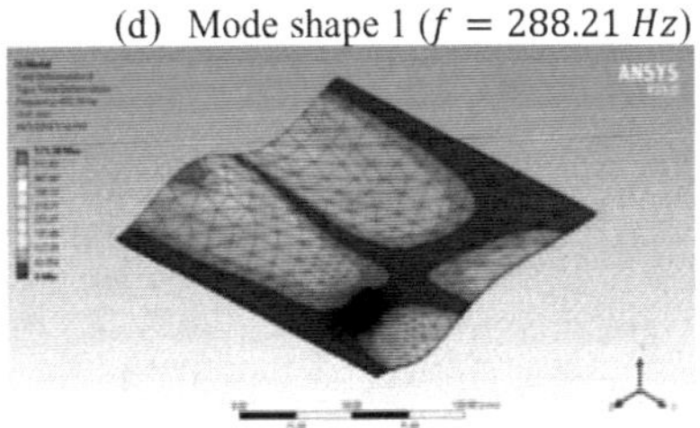

(f) Mode shape 1 ($f = 387.04\ Hz$)

Figure 13. 6 Mode shapes over the range of frequency 3 Hz – 500 Hz

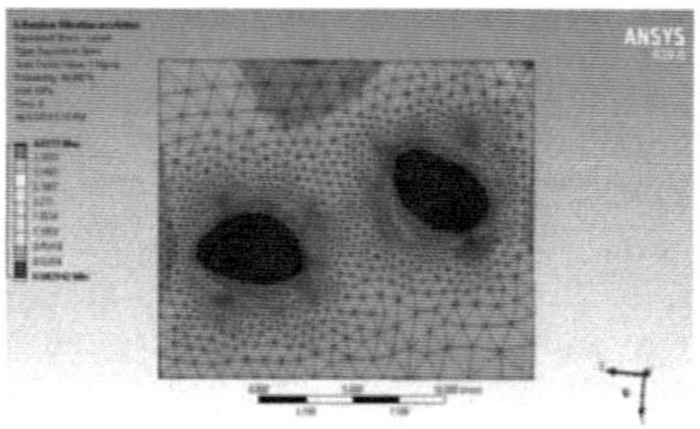

(a) Model 1

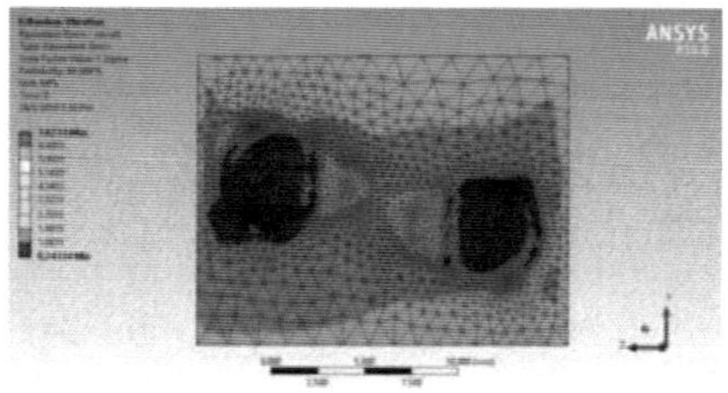

(b) Model 2

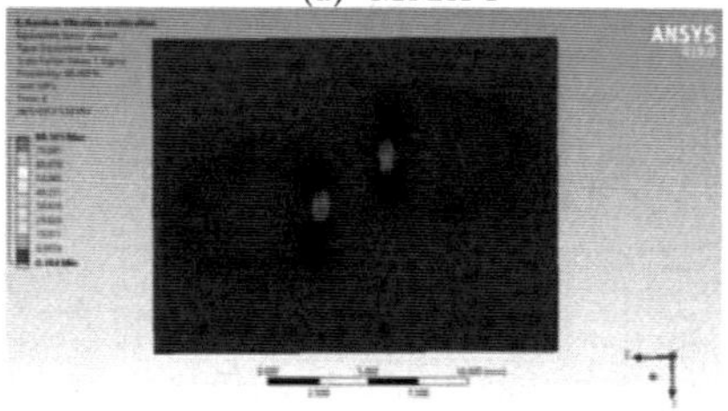

(c) Model 3

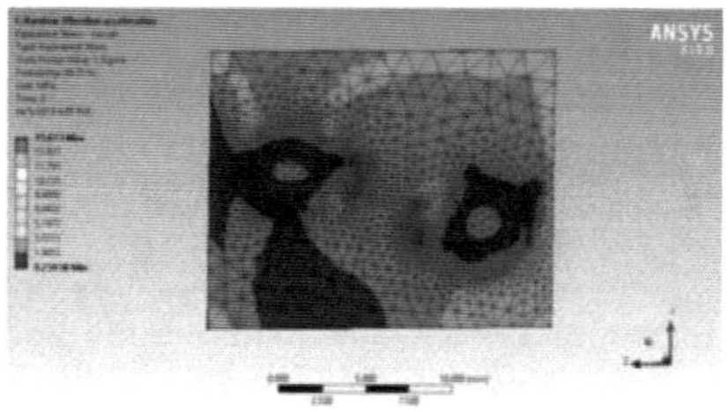

(d) Model 4

Figure 14. Maximum Equivalent stresses contour plot at copper pad circuit

3.5 Random Vibration Analysis

Random vibration is presented in the manner of Gaussian probability distribution with zero – mean and the root mean square (RMS) response is described in the form of sigma values 1σ, 2σ and 3σ, where these deviations lie between RMS stress values over time, with probabilities 68.3%, 95.4% and 99.73% respectively [13].

The maximum equivalent stress occurred at copper pad circuit surface. As shown in figure 14, only model 3 has the stresses at corner of LED that intersects with copper pad while remaining models have different values of stress that are not only affected to intersection with adhesive solder joint but stresses spread on overall; copper pad circuit. Table 4 gives the maximum stress values of each part of LEDs model. It is concluded that model 3 has the highest sigma value and model 1 has the lowest sigma than others.

The stress contour plot obtained in figure 15 for the solder joint shows that stresses are generated at surface below the adhesive joint in contact with the circuit pad. Figure 16 illustrates that the stresses are produced at LED lead where the maximum equivalent stress for model 1 is concentrated at the corners of LED lead and for model 2 the stresses is distributed from the corner of the LED lead to the LED lead base.

The analysis indicated that the most critical areas of all the LED models under vibration loading are at the connections of adhesive joints and circuit pad which means that there are high probabilities that interconnections are cracked or peeled off, either from the copper circuit pad or together with copper circuit pad.

Table 4. Maximum Equivalent stresses (MPa)

Parts	Sigma (σ)	Model 1	Model 2	Model 3	Model 4
copper pad circuit	1σ	4.0215	7.6233	8.8505	5.0244
	2σ	8.0431	15.2470	17.7010	10.0490
	3σ	12.0650	22.8700	26.5510	15.0730
solder joints	1σ	0.6830	1.0485	1.3205	0.6547

	2σ	1.3661	2.0969	2.6410	1.3095
	3σ	2.0492	3.1454	3.9616	1.9643
	1σ	4.7516	8.0057	8.8946	5.6545
LED	2σ	9.5032	16.0110	17.7890	11.3090
	3σ	14.255	24.0170	26.6840	16.9640

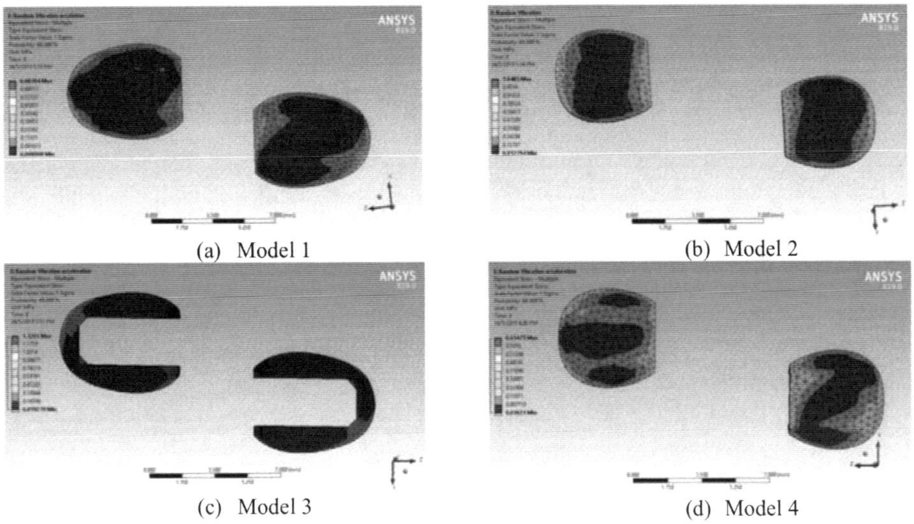

(a) Model 1 (b) Model 2

(c) Model 3 (d) Model 4

Figure 15. Maximum Equivalent stresses contour plot at adhesive solder joints

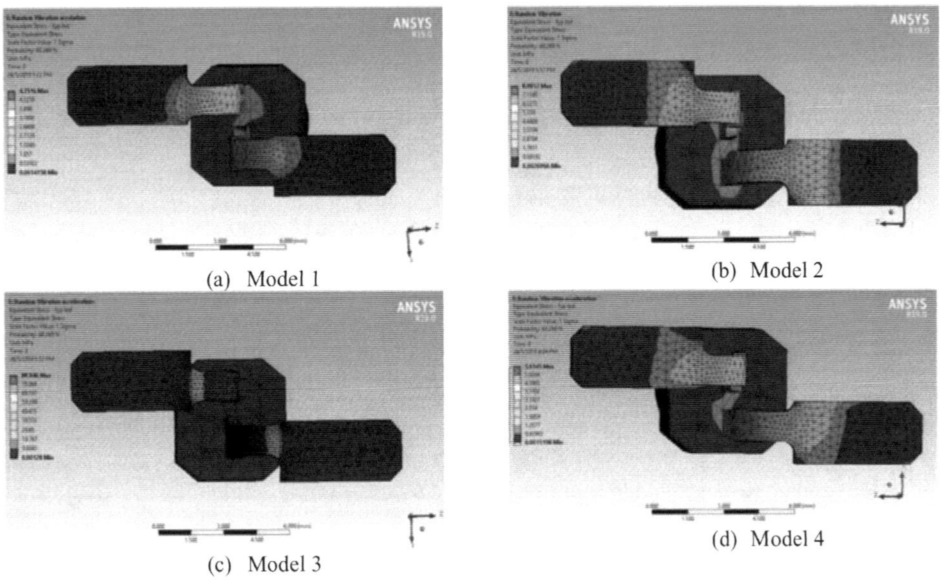

(a) Model 1 (b) Model 2

(c) Model 3 (d) Model 4

Figure 16. Maximum Equivalent stresses contour plot at LED

3.6 Harmonic Response Analysis
Harmonic response analysis is done within the frequency range of 3 Hz – 500 Hz in order to analyse the increasing and decreasing variations of stresses. The variations of stress amplitude under different exciting frequencies for all 4 models are shown in figures 17. It is observed that the stresses are not always increasing but sometimes stresses drop based on a certain frequency. According to this analysis, model 4 shows the highest value and model 2 shows the lowest value of maximum stress amplitude. Therefore, more fractures can occurred in model 4 and cracks are generated on higher maximum stress amplitude.

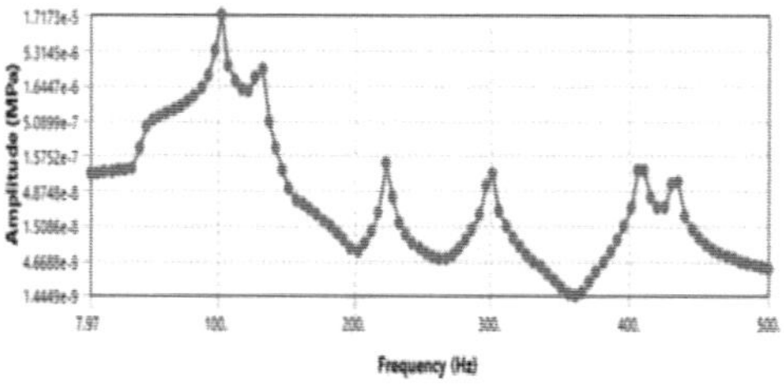

(a) Model 1

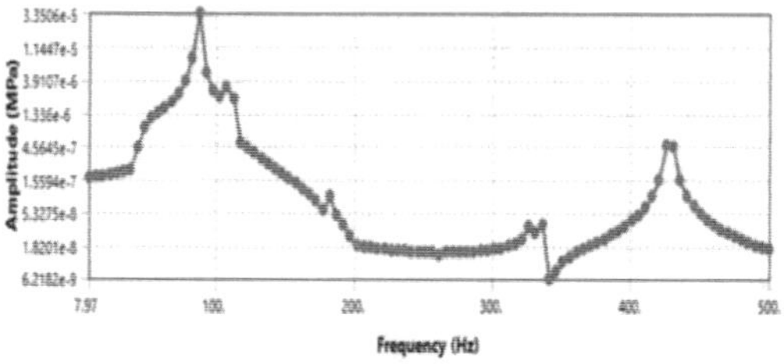

(b) Model 2

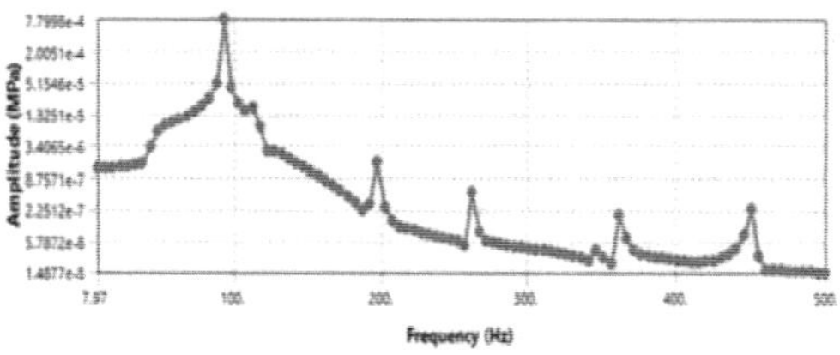

(c) Model 3

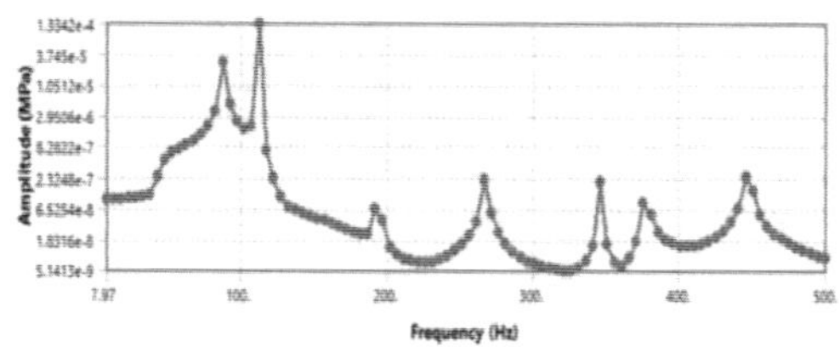

(d) Model 4

Figure 17. Amplitudes for different exciting frequencies

4. Conclusion

The structural reliability of the interconnections on copper pad circuit are determined by the method of dispensing adhesive. Before doing pull strength test analysis, meshing is done with higher number of elements size to achieve accurate results. The circuit pad and LED solder joints proved good structural condition and performance under mechanical loading in pull strength test analysis. As a result, model 3 shows the highest equivalent stress than the others. The solder joints in model 3 can sustained large amount of stress under mechanical and vibration loadings.

Four samples of LEDs adhesive joints can withstand under random vibration within a frequency range of 3 Hz – 500 Hz for 4 hours. After investigation, the critical areas of assembly models are at the contact surface between circuit pads and solder joints. A free vibration analysis also known as harmonic response analysis is done to get more efficient and quick results. This analysis is based on the mode superposition method. According to harmonic response analysis, model 4 shows the highest value and model 2 shows the lowest value of maximum stress amplitude. It is concluded that model 4 has more cracks generated. The stresses generated at each LED joint samples are also supported by experimental results. These results show that the interconnection method can be used in electronic devices because there is no effect on the structural integrity of solder joints under severe vibration conditions.

References

[1] H. C. Pusey, "Book Review : Vibration Analysis For Electronic Equipment (Second Edition) David S. Steinberg John Wiley & Sons New York, Chichester, Brisbane, Toronto, Singapore 1988, 433 pp, $61.95," *Shock Vib. Dig.*, vol. 22, no. 8, pp. 13–13, 1990.
[2] J. F. Rolin, L. Delauney, B. Bigourdan, F. Salvetat, B. J. Galvan, and Y. Elejalde, "Engineering

IOP Conf. Series: Materials Science and Engineering **815** (2020) 012003 doi:10.1088/1757-899X/815/1/012003

approach (reliability analysis, metrology and environmental testing procedures) for newly developed marine sensors," in *OCEANS 2017 - Aberdeen*, 2017.

[3] JEDEC, "Vibration , Variable Frequency," *Jesd22-B103B*, no. July 1989, 2010.

[4] C. M. Ramesha, K. G. Abhijith, A. Singh, A. Raj, and C. S. Naik, "Modal Analysis and Harmonic Response Analysis of a Crankshaft," *Int. J. Emerg. Technol. Adv. Eng.*, vol. 5, no. 6, pp. 323–327, 2015.

[5] C. Chen and H. Yang, "Harmonic resonance analysis for wind-thermal-bundled half-wavelength AC transmission system," *Dianli Zidonghua Shebei/Electric Power Autom. Equip.*, vol. 39, no. 2, 2019.

[6] B. Willis, " Handbook of Lead-Free Soldering Technology for Microelectronic Assemblies20051Karl Puttliz and Kathleen Stalter (Eds). Handbook of Lead-Free Soldering Technology for Microelectronic Assemblies . Marcel Dekker, 1000 pp., ISBN: 0 8247 4870 0 22 sections illustrated with diagrams and photographs ," *Microelectron. Int.*, 2005.

[7] Y. Li and C. P. Wong, "Recent advances of conductive adhesives as a lead-free alternative in electronic packaging: Materials, processing, reliability and applications," *Materials Science and Engineering R: Reports*, vol. 51, no. 1–3. pp. 1–35, 2006.

[8] K. S. Siow and Y. T. Lin, "Identifying the Development State of Sintered Silver (Ag) as a Bonding Material in the Microelectronic Packaging Via a Patent Landscape Study," *J. Electron. Packag. Trans. ASME*, 2016.

[9] R. A. Amy, G. S. Aglietti, and G. Richardson, "Reliability analysis of electronic equipment subjected to shock and vibration - A review," *Shock and Vibration*. 2009.

[10] Y. S. Chen, C. S. Wang, and Y. J. Yang, "Combining vibration test with finite element analysis for the fatigue life estimation of PBGA components," *Microelectron. Reliab.*, 2008.

[11] D. Yu, A. Al-Yafawi, T. T. Nguyen, S. Park, and S. Chung, "High-cycle fatigue life prediction for Pb-free BGA under random vibration loading," *Microelectron. Reliab.*, 2011.

[12] J. B. Libot, L. Arnaud, O. Dalverny, J. Alexis, P. Milesi, and F. Dulondel, "Mechanical fatigue assessment of SAC305 solder joints under harmonic vibrations," in *2016 International Conference on Electronics Packaging, ICEP 2016*, 2016.

[13] S. Kumar, "Analyzing Random Vibration Fatigue," *ANSYS Advant.*, 2008.

Acknowledgements

The authors appreciates the support of Universiti Sains Malaysia through Bridging Grant 304.PMEKANIK.6316085 and Research University Grant 1001.PMEKANIK.8014067. Also, assistance from ETS Department of Jabil Penang, The VibrationLab USM and Advanced Packaging & SMT Unit of USM are acknowledged.

IOP Conf. Series: Materials Science and Engineering **815** (2020) 012004 doi:10.1088/1757-899X/815/1/012004

Sound and Vibration Damping Properties of Water Retted Kenaf Bast Fiber (WKF) and Poly (Butylene Succinate) PBS

WMA Wan Mamat Ali[1]*, L.E. Ooi[1], And M.Z.A Thirmizir[2]

[1]TheVibrationLab, School of Mechanical Engineering, Universiti Sains Malaysia,
[2]Science and Engineering Research Centre, Engineering Campus, Universiti Sains Malaysia.

*Corresponding author: amri@usm.my

Abstract. Nowadays, considerable attention has been paid on the utilization of natural materials to reduce the carbon emission from material synthesis. In this study, the sound and vibration damping properties of natural material based Poly Butylene Succinate(PBS) and Water Retted Kenaf Bast Fiber(WKF) were explored and characterized. The goal is to characterize the sound absorption and damping properties of the PBS and WKF at difference materials percentage. The (material)PBS and WKF are produced by using Compression moulding machine model Gotech GT-7014-H and the sample thickness is 3 mm. The vibration damping measurement of the material is carried out based on the free vibration method technique. It is observed that the damping factor is obtained is 0.0469 for PBS material. The measurement of the sound absorption coefficient of the PBS and WKF is carried out using the impedance tube. The result shows that the maximum sound absorption of 0.305 was observed at a frequency of 642 Hz for PBS material. Whereas Tensile Strength, for PBS material shows 26.33 MPa. These results suggest that the PBS could be a viable candidate for applications which need good sound and vibration properties.

1. Introduction

Noise pollution has become a new pollution source that caused the bad influence on the environment and human health. Nowadays, researchers have carried out various noise control methods for investigating the issue such as passive and active control [1]. The research focus is mostly on natural fibers such as kenaf, ramie and jute reinforced composites since this particular material have a good mechanical property, lightweight, environmentally friendly and biodegradable [2,3].

Multiple layered coir fiber absorbers involving air spaces and micro-perforated plate has been studied both through simulation and experiments [4-5]. Ersoy and Kucu [6] studied the absorption performance of industrial waste tea leaf fiber and found that the absorption coefficient of the fibers increases significantly when backed with a single woven cotton cloth. Yang and Li [1] studied three types of natural fibers which are ramie, flax and jute. They found that those natural composites had satisfying noise reduction ability similar to that of the synthetic fibers. Bastos et al., [7] proposed noise control panels which made from vegetable fibers consisting of coconut, palm, sisal and acai. The performance of vegetable fibers could be greater in certain cases when compared to synthetic acoustic foam materials. Besides, Xiang et al., [8] studied the sound absorption properties of kapok fibers where the result showed that kapok fibers have similar acoustics performance with the combinations of commercial glass wool and cotton fibers although the bulk density of kapok fiber is lower.

Sugarcane waste fibers was studied by Putra et al. [9]. In research found that the sound absorption coefficient is comparable with those of commercial synthetic fibers. The review of previous research shows that there have been many studies performed on natural material that resulted good sound-absorbing and vibration damping properties [10]. These natural materials could essentially be grown for manufacturing composites, in turn providing such benefits as being both biodegradable and recyclable.

This study aimed to explore the mechanical properties, sound absorption coefficient ability and damping ratio of water retted kenaf bast fiber and polybutylene succinate.

2. Experimental Measurement

2.1. Material preparation

Polybutylene succinate (PBS) with trade name Bionolle #1020 is produced by Showa Highpolymer Co. Ltd, Japan. The density of this polymer is 1.25 g/cm^3 and the melt flow index is 26 g/10min (at 190°C and 2.16 kg load). PBS was supplied in dry pellet and has a melting temperature (Tm) ranging from 120-140 °C. Water retted kenaf bast fiber was supplied by Lembaga Kenaf & Tembakau Negara (LKTN). The fiber length distribution ranging from 1-3mm. The fiber density is 1.14 g/cm^3.

2.2. Preparation of the composites

Initially, WKF and PBS were dried in an oven at 50°C for 24 hours and kept in desiccators. Next, the composites were mixed using rotating double-winged rotors (POTOP). The temperature of the compounding chamber was set at 165°C. Generally, the temperature of the chamber was allowed to stabilize for approximately 3 minutes before it was loaded with PBS pellets. Then, the fiber was gradually added to the chamber under a low rotor speed (10 rpm). Once all of the fiber was fed in, the compounding process was continued at a rotor speed of 50 rpm for 5 minutes before being discharged. The composite compounds were compression-moulded at 165°C into dumbbell shape specimens ASTM D638 type I. with a compression-moulding machine Gotech (China). Figure 1 shows the compression mold machine that used in this research. The moulding cycle involved 5 min of preheating. without pressure, 3 min of compression under 50-ton compressive load, and 5 min of cooling under the same load. Before testing, the moulded specimens were kept in desiccators and conditioned at room temperature for at least 24 h. Table 1 shows the details formulation of the composites.

Table 1. The details formulation of the composites.

Designation	PBS (wt. %)	WKF (wt. %)	Weight (gram)
PBS	100	0	3.25
90PBS10WKF	90	10	3.73
80PBS20WKF	80	20	3.6
70PBS30WKF	70	30	3.48
60PBS40WKF	60	40	3.36

Figure 1. Compression molding machine model Gotech GT-7014-H

2.3. Sound absorption coefficient measurement

The Sound Absorption Coefficient of each sample was measured by utilizing of the impendent tube, LMS Scadas Mobile, LMS Software (LMS Sound Absorption Coefficient) and two Microphone units (Grass 46AE) as indicated in figure 2. The plunger at the back of the impedance tube is functioned to control the air cavity gap. The air cavity was set 50 mm for this experiment. A loudspeaker is used as a sound source on the impendent tube and it is placed in front of the impendent tube as shown in figure 2. It generates the broadband random sound waves and the sound waves propagating as plane waves in the tube hit the sample, get partially absorbed, and subsequently reflected. The acoustical properties of the test sample were tested in the frequency range of 50–1550 Hz. For this experiment the size of each sample was 34.8mm diameter and 3mm thick. The detail of the samples used with a different percentages of PBS and WKF is shown as in figure 3.

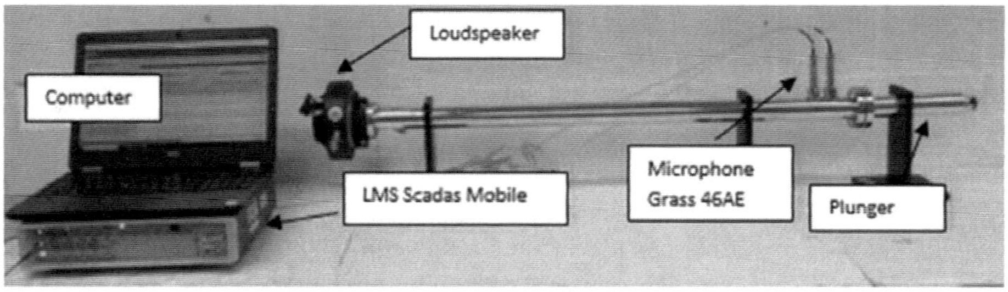

Figure 2. The setup experiment of the Sound Absorption Coefficient.

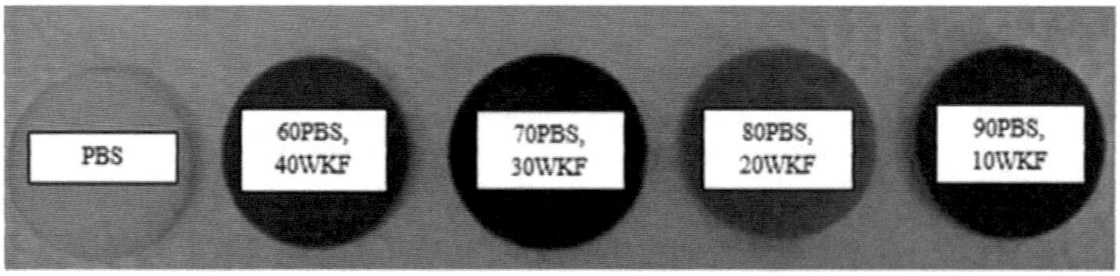

Figure 3. Sound Absorption Coefficient test Specimens.

2.4. Vibration damping factor measurement.

Free vibration method is used to measure damping factor as shown in figure 4. The specimens were cut to desired size of 165mm x 12mm x 3mm. In figure 5, a different specimen of the percentage of PBS & WKF is used for damping factor measurement. The specimen was mounted on a cantilever beam structure and fastened using a fixture. An accelerometer (Dytran, 3224A2) with a sensitivity of 2.56 mV/g mounted at the end of the specimen. Before measurement, each accelerometer was calibrated using vibration calibrator. The signal output was process by LMS Scadas mobile data acquisition linked with LMS Test Xpress software on the personal PC.

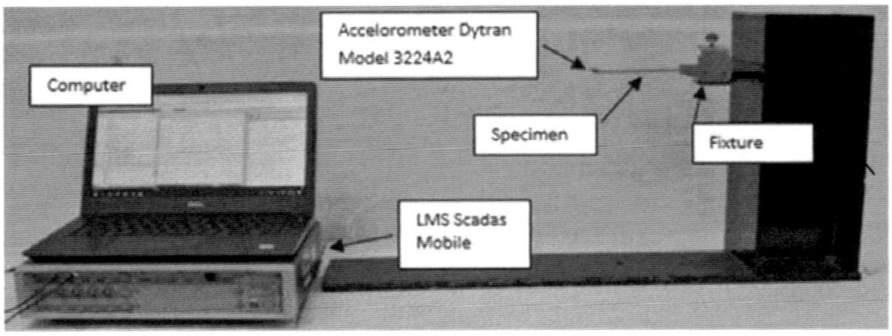

Figure 4. Vibration damping test set up

Figure 5. Vibration damping test specimen.

Figure 6 shows the vibration response of the specimen over time. The damping factor is calculated by taking the highest value and the lowest value at the top of the graph in the obtained waveform. Then take a few peaks of the graph to calculate the damping factor (ζ) and the obtained values will be included in equations 1 and 2.

$$\delta = \frac{1}{\eta} \ln \left(\frac{y_0}{y_n}\right) \tag{1}$$

$$\zeta = \frac{1}{\sqrt{1 + \left(\frac{2\pi}{\delta}\right)^2}} \tag{2}$$

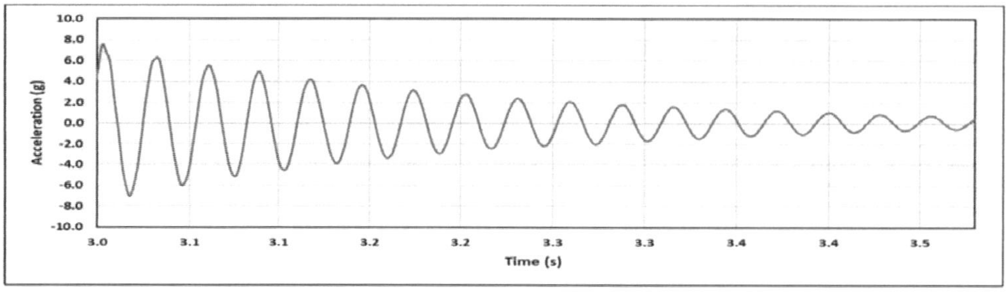

Figure 6. Free Vibration response of the PBS & WKF.

2.5. Tensile measurement Set Up

The tensile strength, elongation and Young's modulus and of the proposed composites were determined using universal testing machine model Instron 5969 (USA) at room temperature and 50% relative humidity. The dumbbell-shaped samples in accordance with ASTM D638 (Type I) with a thickness of 3 mm and width 12.7 mm have been tested at an extension rate of 5 mm/min.

3. Result and discussions

3.1. Analysis of sound absorption coefficient measurement

Figure 7 shows the results of the sound absorption coefficient (SAC) of materials with different percentages of PBS and WKF at 3mm thickness. The graph shows that the SAC of purely PBS is higher compare to the composite material which is 0.305. The peak sound absorption was observed at 642 Hz and for all materials. The SAC of 60PBS 40WKF is 0.263, 0.257 for 70PBS 30WKF, 0.278 for 80PBS 20WKF and 0.147 for 90PBS 10WKF material. From all five advanced composite materials, PBS with higher sound absorption coefficient can be used accurately in sound absorption applications as an important design criterion. Besides, the mixing rate between PBS and WKF also affects the results produced by the Sound Absorption Coefficient.

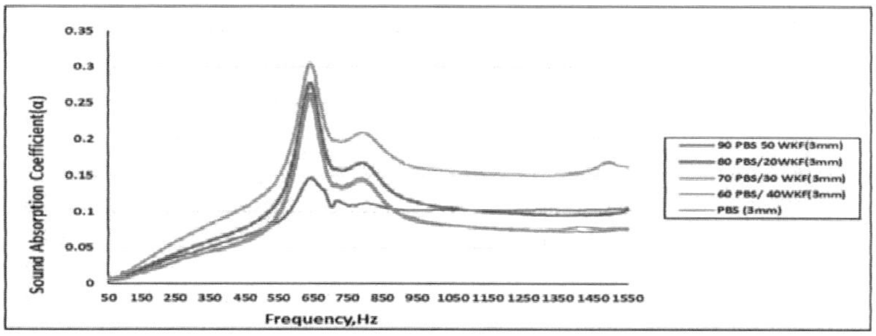

Figure 7. Sound Absorption Coefficient for specimen at 3 mm.

3.2. Analysis of vibration damping factor measurement.

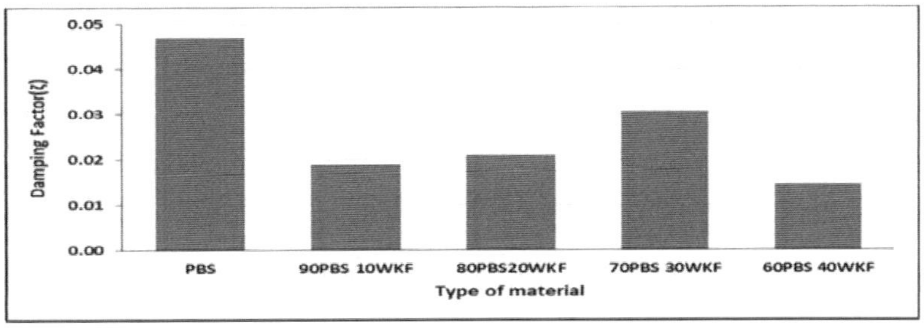

Figure 8. Damping ratio at different material, for thickness 3 mm.

Figure 8 shows the result of the damping ratio for specimen thickness of 3 mm for material with different percentages of PBS and WKF. From the results, the damping ratio for purely PBS material was 0.0469. Whereas the damping ratio for 60PBS 40WKF material is 0.0145, 0.0305 for 70PBS 30WKF, 0.021 for 80PBS 20WKFand 0.019 for 90PBS 10WKF. The results show that the PBS material has the highest damping ratio compared to the others.

3.3. Analysis of Tensile strength measurement.

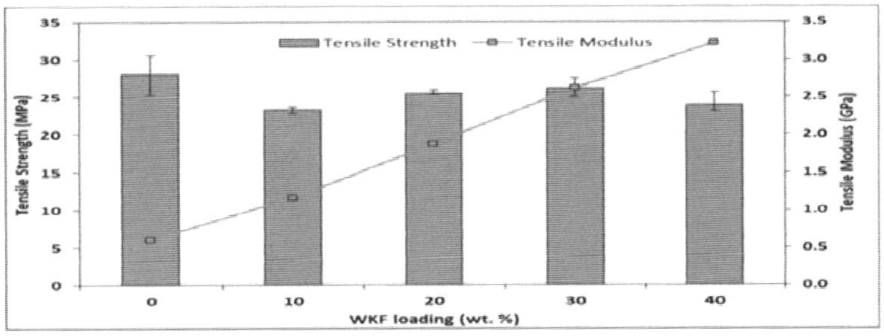

Figure 9. Tensile strength and modulus of PBS/WKF composites at different fiber loading.

Figure 9: shows the effect of fiber loading on the tensile strength and modulus of PBS/WKF composites. In general, the neat PBS exhibits higher tensile strength than the PBS/WKF composites. The poor tensile strength of the composites especially at lower fiber loading i.e. 10 and 20 wt. % was due to poor fiber dispersion in the matrix as a result of incompatible fiber-matrix interaction. As already known, PBS is a hydrophobic polymer and kenaf fiber is hydrophilic and the incorporation of both materials in composites will result in poor interfacial adhesion and fiber-matrix interaction by Yatigala et al., [11]. However, the composites show increases of tensile strength from 23.34 MPa, to 25.56 MPa and 26.13 MPa when fiber loading increased from 10, 20 and 30 wt. %. The increases in tensile strength could be due to the reinforcing effect the WKF and also physical fiber-to-fiber entanglement of the kenaf fibers by Ismail & Ishak, . [12] The optimum fiber loading was at 30 wt. % and an abrupt drop was observed when the fiber loading further increased to 40 wt. %. The reduction of tensile strength at higher fiber loading was due to severe fiber agglomeration as a result of insufficient wetting of fiber by the polymer matrix by Han et al., [13]. The fiber agglomerates will be a stress concentration points and will lead to premature failure of the composites. On the other hand, the addition of kenaf also increased the tensile modulus of the composites because of the stiff nature of the WKF. The higher the fiber loading the higher tensile modulus of the composites. The consistent increment in the tensile modulus with fiber loading was probably due to the restriction of the polymer matrix mobility with the presence of rigid fibers hence make the composite stiffer by Hamma et al., [14]. As a consequence, the elongation at break of the composites was also reduced with increasing fiber loading as shown in Figure 10. As will-known, the elongation at break property mainly influenced by the ductility of the matrix phase which controls the deformation behaviour of the composites. The reduction of matrix portion or the restriction of its deformability will alter the deformation of the composite.

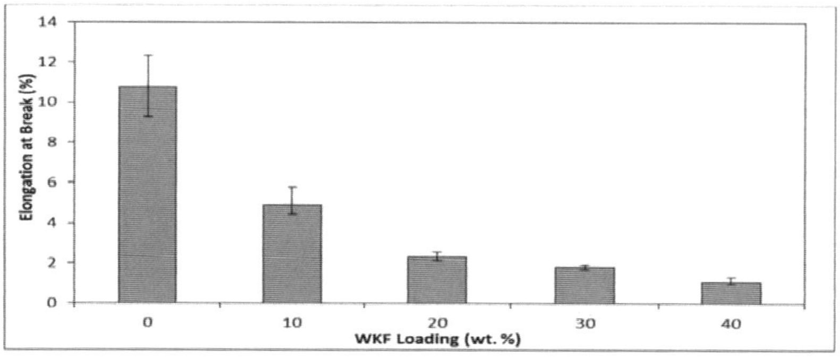

Figure 10. Elongation at break of PBS/WKF composites at different fiber loading.

4. Conclusion

In this study, the effect of mixing the percentage difference between PBS and WKF on the sound absorption coefficient was investigated. Based on the results, when the PBS and WKF materials were mixed with different percentages it showed that the noise absorption had decreased. Similarly, the Damping ratio test and tensile strength were found to decrease. According to the experimental data, the PBS material was better than the mixture between PBS-WKF.

5. References

[1] W. Yang and Y. Li, "Sound absorption performance of natural fibers and their composites," *Sci. China Technol. Sci.*, vol. 55, no. 8, pp. 2278–2283, 2012.

[2] A. K. Mohanty, M. Misra, and L. T. Drzal, "Surface modifications of natural fibers and performance of the resulting biocomposites: an overview," *Compos. interfaces*, vol. 8, no. 5, pp. 313–343, 2001.

[3] H. Cheung, M. Ho, K. Lau, F. Cardona, and D. Hui, "Natural fibre-reinforced composites for bioengineering and environmental engineering applications,"
Hamma, A., Kaci, M., Mohd Ishak, Z., Ceccato, R., & Pegoretti, A. (2015). Starch-grafted-polypropylene/kenaf fibres composites. Part 2: thermal stability and dynamic-mechanical response. *Journal of Reinforced Plastics and Composites*, *34*(24), 2045–2058. https://doi.org/10.1177/0731684415609792.

[4] M. J. M. Nor, N. Jamaludin, and F. M. Tamiri, "Effect of perforated panel in acoustics absorption performance of coir fiber material," in *12th International Congress on Sound and Vibration 2005, ICSV 2005*, 2005.

[5] M. J. M. Nor, N. Jamaludin, and F. M. Tamiri, "A preliminary study of sound absorption using multi-layer coconut coir fibers," *Electron. J. Tech. Acoust.*, vol. 3, pp. 1–8, 2004.

[6] S. Ersoy and H. Küçük, "Investigation of industrial tea-leaf-fibre waste material for its sound absorption properties," *Appl. Acoust.*, vol. 70, no. 1, pp. 215–220, 2009.

[7] L. P. Bastos, G. da S. V. de Melo, and N. S. Soeiro, "Panels manufactured from vegetable fibers: An alternative approach for controlling noises in indoor environments," *Adv. Acoust. Vib.*, vol. 2012, 2012.

[8] H. Xiang, D. Wang, H. Liua, N. Zhao, and J. Xu, "Investigation on sound absorption properties of kapok fibers," *Chinese J. Polym. Sci.*, vol. 31, no. 3, pp. 521–529, 2013.

[9] A. Putra, Y. Abdullah, H. Efendy, W. M. Farid, M. R. Ayob, and M. S. Py, "Utilizing sugarcane wasted fibers as a sustainable acoustic absorber," *Procedia Eng.*, vol. 53, pp. 632–638, 2013.

[10] S. Prabhakaran, V. Krishnaraj, and R. Zitoune, "Sound and vibration damping properties of flax fiber reinforced composites," *Procedia Eng.*, vol. 97, pp. 573–581, 2014.

[11] Yatigala, N. S., Bajwa, D. S., & Bajwa, S. G. (2018). Compatibilization Improves Performance of Biodegradable Biopolymer Composites Without Affecting UV Weathering Characteristics. *Journal of Polymers and the Environment*, *26*(11), 4188–4200. https://doi.org/10.1007/s10924-018-1291-7.

[12] Ismail, N. I., & Ishak, Z. A. M. (2018). Effect of fiber loading on mechanical and water absorption capacity of Polylactic acid/Polyhydroxybutyrate-co-hydroxyhexanoate/Kenaf composite. *IOP Conference Series: Materials Science and Engineering*, *368*, 012014. https://doi.org/10.1088/1757-899X/368/1/012014.

[13] Han, S. O., Karevan, M., Sim, I. N., Bhuiyan, M. a., Jang, Y. H., Ghaffar, J., & Kalaitzidou, K. (2012). Understanding the Reinforcing Mechanisms in Kenaf Fiber/PLA and Kenaf Fiber/PP Composites: A Comparative Study. *International Journal of Polymer Science*, *2012*, 1–8. https://doi.org/10.1155/2012/679252.

[14] Hamma, A., Kaci, M., Mohd Ishak, Z., Ceccato, R., & Pegoretti, A. (2015). Starch- grafted-polypropylene/kenaf fibres composites. Part 2: thermal stability and dynamic-mechanical response. *Journal of Reinforced Plastics and Composites*, *34*(24), 2045–2058. https://doi.org/10.1177/0731684415609792

IOP Conf. Series: Materials Science and Engineering **815** (2020) 012005 doi:10.1088/1757-899X/815/1/012005

Monitoring of Gear Surface Roughness in Angle Grinder using Vibration Analysis

Z N M Yusuf[1]*, N S D Sidek[1], M N A Hamid[1], P Y Teoh[2]

[1]Universiti Kuala Lumpur Malaysian Spanish Institute, Kulim Hi-Tech Park, 09000 Kulim, Kedah, Malaysia

[2]Keyence Sdn. Bhd, Malaysia

*Corresponding author: zainalnazri@unikl.edu.my

Abstract. This paper aims to provide the relation between surface roughness of angle grinder towards vibration amplitude. Prolong usage of angle grinder over excessive force will affect the performance of angle grinder where it tends to vibrate due to surface roughness of gear and create excessive vibration that will lead to hand-arm vibration system. Therefore, this study is conducted to investigate surface roughness of the gear teeth of angle grinder affects the amplitude of vibration. The amplitude of the vibration for angle grinder is conducted by using LMS data acquisition system. Meanwhile, the roughness of the gear teeth is measured by using Keyence Laser Confocal Microscope. The data was taken by every 100 hours of grinding process. The results show that the surface roughness of the gear teeth is steadily decreasing due to contact between gears and subsequently resulted increasing the amplitude of the vibration. For 500 hours measurement, the correlation of surface roughness of gear teeth with the amplitude of vibration is 93.7% for marked tooth and 86.5% for adjacent tooth, this indicates the strongest possible agreement that when surface roughness of the gear teeth decreasing, the vibration amplitude is higher.

1. Introduction

This project is specifically focusing on gear's roughness in the angle grinder; bevel gear. Rotating machinery produces vibration during its normal operation as a consequence of friction and centrifugal forces of both the rotating parts and the bearings. The contacts of disc grinder with the workpiece generate excitation force that can affect the component such as gear, motor and bearing assembly. With frequent use of angle grinder in fabrication activities, the gears interaction will result in surface roughness and subsequently generate high level of vibration. The effects of the friction toward the gear assembly produce surface roughness of the gear over the length of time used. Monitoring the gears roughness is very important to determine the tendency of high vibration level in grinder. Trend analysis is used to analyse the relation of surface roughness and vibration amplitude. Trend monitoring of vibration level is a more useful maintenance tool than a one-time survey of the absolute magnitudes of features only. A slight linear increase of feature values turns to exponential when the point of failure is approaching. The time of failure depends strongly on machines and the stress caused by operating conditions. The paper is divided into different section, each dealing with various aspects of the subjects: it begins with a summary of review of vibration analysis, surface roughness measurement and followed by general overview of gear. The gear monitoring technique in angle grinder is discussed in detail which is vibration analysis. One of the major areas of interest in the modern-day condition monitoring of rotating machinery is that of vibration. If a fault developed and goes undetected, then, at best, the problem will not be too serious and can be remedied quickly and cheaply; at worst, it may result in expensive damage and down- time, injury, or even loss of life. By measurement and analysis of the vibration of rotating machinery, it is possible to detect and locate important faults such as mass unbalance, gear faults, misalignment, and crack.

2. Literature Review

The aim of this section is to provide the reader with an understanding of the gear system, vibration analysis and gear surface roughness measurement.

2.1. Vibration Monitoring Methods

Vibration monitoring is one of the main tools that allow determining the mechanical health of various components of a machine in a non-intrusive manner. Reducing the vibration is important because it shows that smaller vibration amplitude and frequency will reflect better surface and surface quality [1]. The mechanical components that are the most often encountered in rotating machinery are bearings - roller bearings in particular – and gears. In addition, these components are generally the most loaded and therefore subject to early damage in the life of the machine [2]. Vibration-based damage detection for rotating machinery (RM) has been repeatedly applied with success to a variety of machinery elements such as roller bearings and gears.

2.1.1. Spectral Analysis

The vibration signature character in the frequency domain is most prominent. That is why many control methods are focused on the vibration signature's frequency domain characteristics, the continuum. The most successful spectral analysis is applied to the low and intermediate frequency range. The strategy is to extract the signature from a single gear or bearing from the total vibration measured on the machine casing and to monitor the characteristic defect frequency amplitudes. The basic assumption; that a growing defect implies increasing amplitudes at the defect frequencies. Some defect indicators, like the defect severity index, focus on the difference in amplitude at the defect frequency and the amplitude of the background. Other alternative methods focus on the side bands and attempt to build indicators that measure the relative amplitudes of the side bands [3].

2.1.2. Gear Fault

The vibrations of a gear are mainly produced by the shock between the teeth of the two wheels. Gear fault is simulated with filled between teeth. The vibration monitored on a faulty gear generally exhibits a significant level of vibration at the GMF tooth meshing frequency (i.e. the number of teeth on a gear multiplied by its rotational speed) and its harmonics, the distance of which is equal to each wheel's rotational speed [4]. Gear faults can be generally classified into two major categories: distributed faults and local faults. Distributed faults are those faults that results from poor gear mounting, or manufacturing inaccuracies such as eccentricities, varying gear tooth spacing, etc. Meanwhile, local faults are those caused by localizing defects that can occur in gear teeth such as wear of the tooth surface, cracks in gear teeth, and loss of part of the tooth due to breakage or loss of the entire teeth [5].

2.1.3. Correlation between Roughness and Vibration Amplitude

The evaluation of the vibration signals observed during grinding wheel wear in an attempt to determine the tool life. His study determines the average roughness of the work piece increases with the material removed and thus the grinding wheel wear. This in turn results in higher vibration amplitude seen on the tailstock. It concludes by claiming the claimed tool life can be improved with a better signal conditioning algorithm [6]. All gear operations involve some form of sliding contact between the mating gear surfaces. This sliding contact can produce random vibrations from the asperity contacts between the gears, the nature of which is dependent on a number of factors, such as the speed and load, the lubrication conditions and the micro-geometry of the surfaces. Changes in surface roughness would no doubt affect the overall vibration (e.g., rms) level and its frequency characteristics, but it would be very difficult to distinguish surface roughness from other effects [7].

3. Methodology

The aim of this section is to provide the reader about the method that has been done in this paper. The experiments are conducted for after every 100 hours of grinding process. After the grinding process, collection of data will be done; vibration data & roughness measurement data. The correlation data between surface roughness and vibration amplitude has been done using statistical tool which is Minitab19.

3.1. Flowchart

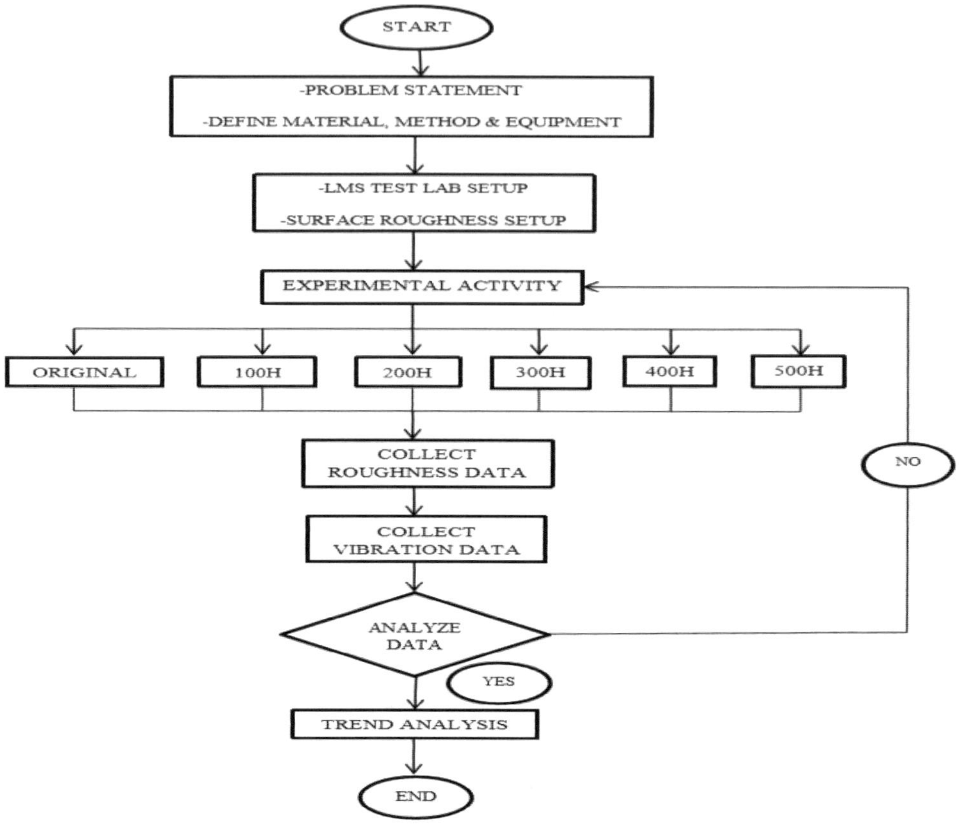

Figure 1. Process Flow

3.2. Gear Specification

Power transmission using bevel gear mechanism can be described as in figure 2. The gear ratio helps to determine the output and input of the speed as well as the torque and power. The gear ratio can be calculated using equation (1). It is vital to determine the gear ratio respect to the individual gear and pinion rotating speed to determine gear mesh frequency in order to locate the appropriate frequency band as given in equation (2).

Table 1. Gear Specification

Number of teeth on driven gear	36
Number of teeth on driver gear	11
Gear Ratio	3.27

Figure 2. Bevel Gear

$$G.R = \frac{T2}{T1} \quad (1)$$

GR = Gear Ratio
T1 = Number of teeth on driven gear
T2 = Number of teeth on driver gear

$$G.M.F = \frac{T.N}{60} \quad (2)$$

GMF= Gear Mesh Frequency (Hz)
T = Number of Gear Teeth
N = Shaft Speed (rpm)
The numbers of gear teeth are 36 and the RPM are 10805. The RPM measured from output gear.

3.3. LMS Data Acquisition
LMS Data acquisition system with 8 channels inputs was used to measure the vibration amplitude of angle grinder. The spectral test lab software was used to filter the data obtained and processed from time domain to frequency domain technique. The procedures such as channel setup, calibration and scope setting were carried out in advance to make sure all data collected are accurate. The detailed scope settings are tabulated below.

Table 2. Scope Settings

Bandwidth	8192.00 Hz
Spectral Lines	8192
Resolution	1.0 Hz

After the setup has been done, the accelerometer then was properly attached to the bearing on the angle grinder to collect the vibration data.

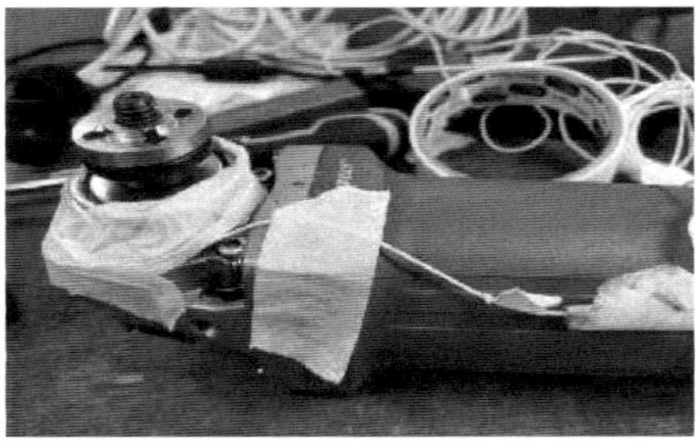

Figure 3. Location of accelerometer

3.4. Roughness Measurement using Keyence Laser Confocal Microscope

For the surface roughness measurement, two tooth surfaces of the driving gear labelled as marked tooth and adjacent tooth is measured before and after each cycle. Measurement of surface roughness is conducted using the Keyence Laser Microscope. The profile surface features can be seen with details such as the peaks and valleys by using the magnifying lens (20x). These peaks and details created by the grinding process can be seen clearly along the axial direction of gears. This instrument is calibrated before surface roughness of the tested tooth is taken. The calibration is performed to ensure that high precision of surface roughness measurement is maintained. The processes start with optical view followed by laser view.

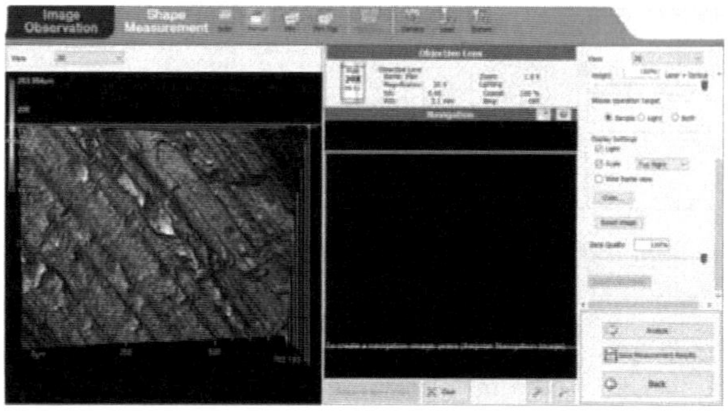

Figure 4. Optical View for Image Observation

The average surface roughness (Ra) is used to observe changes in the tooth surface roughness.

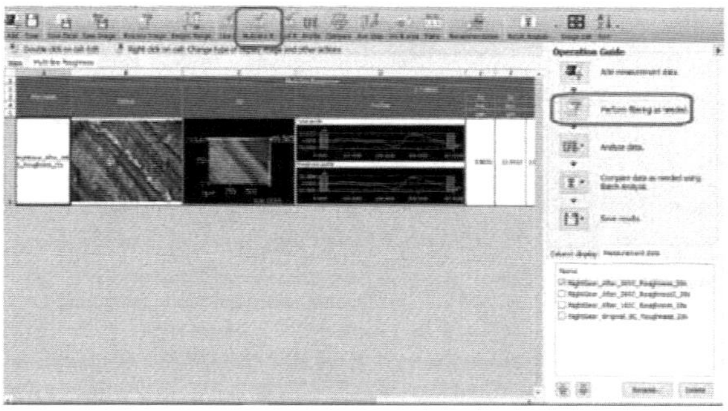

Figure 5. Process of Roughness Measurement

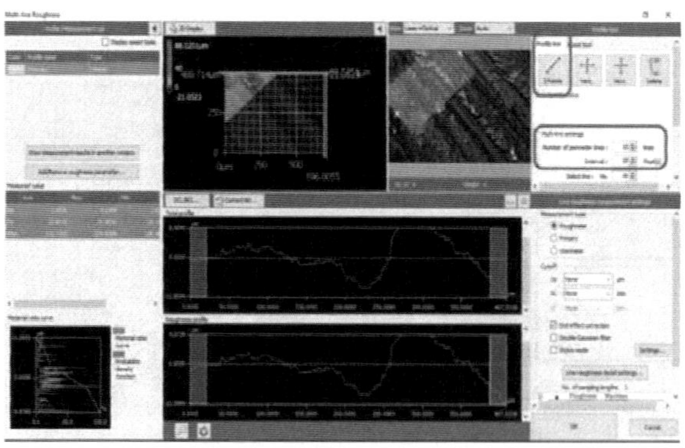

Figure 6. Multi-line Settings

Table 3. Roughness Setting

Resolution	1024x768
Perimeter Lines	15 lines
Interval	10 pixels

4. Result and discussion

In this section, the results are presented, and the behavioural analysis is explained in detail.

4.1. Vibration Data

The frequency of vibration data obtained through the collection of data from LMS. The measurement system is set with chosen frequency which is 6483 Hz. The vibration measurements show that the amplitude steadily increasing after every cycle run. The comparisons between the values of vibration data illustrate that amplitude trend increasing after each cycle run.

Table 4. Vibration Data

Operating Hour(s)	Amplitude Data
0 (baseline)	0.20
100	0.44
200	0.62
300	0.86
400	0.90
500	1.00

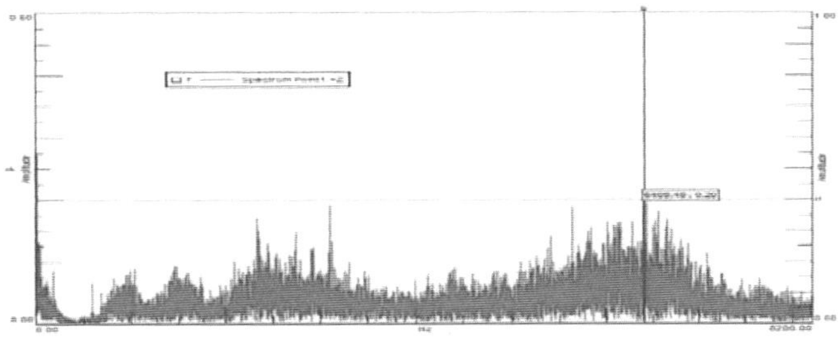

Figure 7. Vibration Amplitude Original

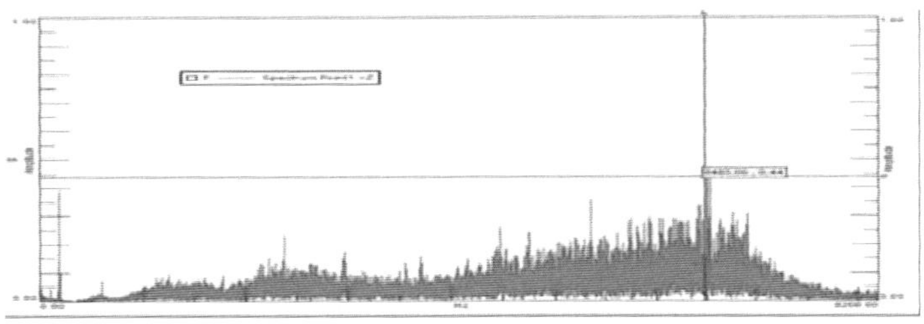

Figure 8. Vibration Amplitude 100 Hours

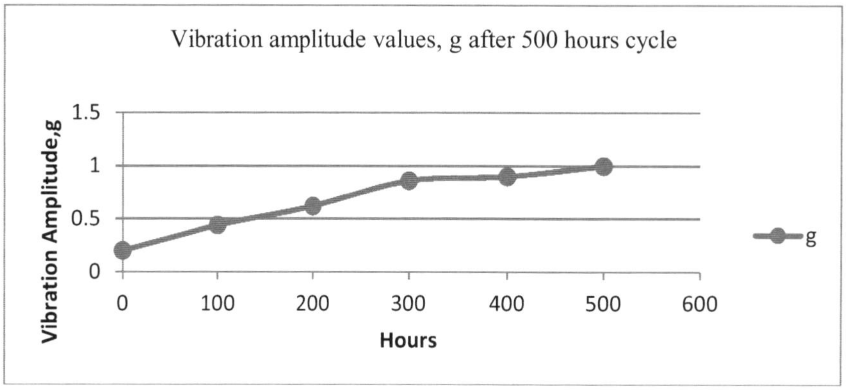

Figure 9. Vibration Amplitude after 500 Hours

From these results, the vibration amplitude showed an increasing trend due to gear tooth wear that lead to backlash of gear.

4.2. Surface Roughness Data

The surface roughness measurements, which are obtained through laser microscope of bevel gear in the radial direction, facilitate detailed observations of features of contact surfaces such as the peaks. There two teeth of gear that has been measured and labelled marked tooth and adjacent tooth. The surface roughness measurements show that the tooth surface roughness is changed after every cycle run. The comparisons between the values of surface roughness illustrate that the peak values of surface roughness deteriorate after each cycle run. The peaks decrease from 7.339 to 2.6804 μm for marked tooth while for adjacent tooth, the peak decreases from 6.404 to 3.5422 μm. The surface roughness of gear tooth (Ra) has a steady trend of decrease, as figure 10 and figure 11 shows.

Table 5. Profile of Roughness

	300 HOURS	400 HOURS
MARKED TOOTH	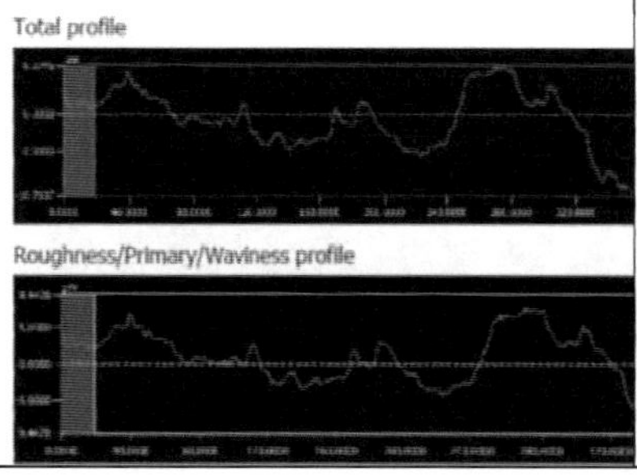	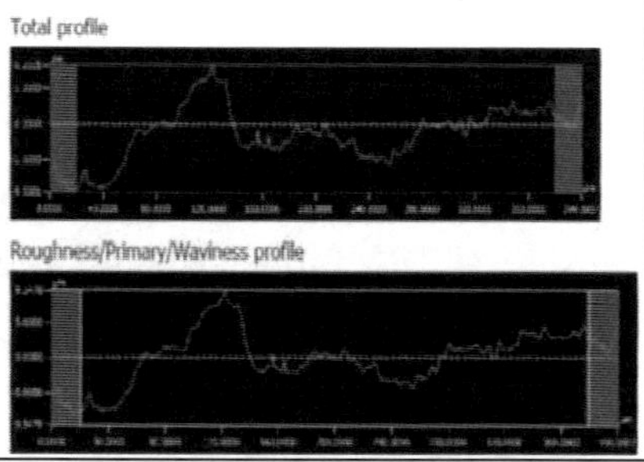

IOP Conf. Series: Materials Science and Engineering **815** (2020) 012005 doi:10.1088/1757-899X/815/1/012005

ADJACENT TOOTH	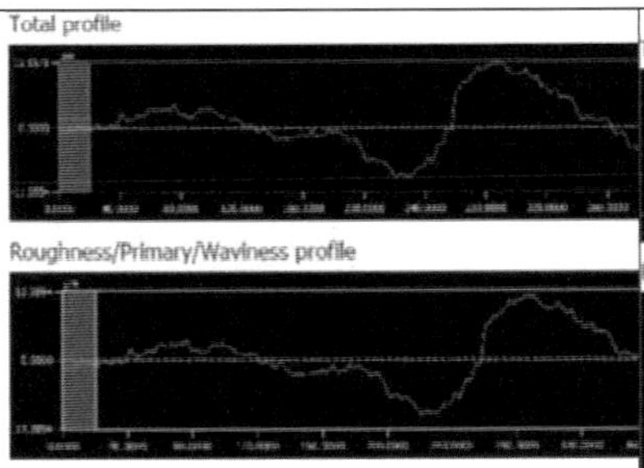 	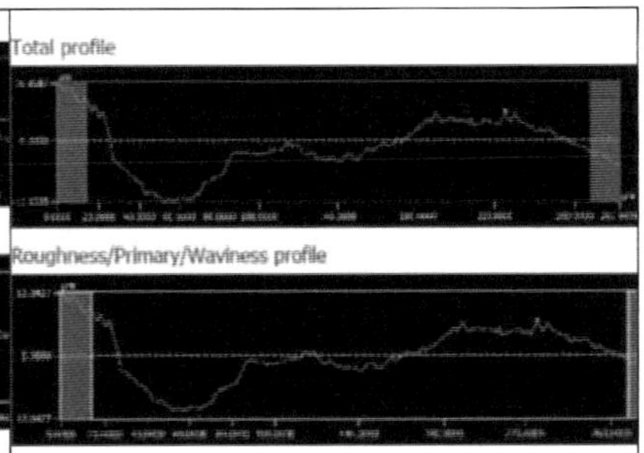

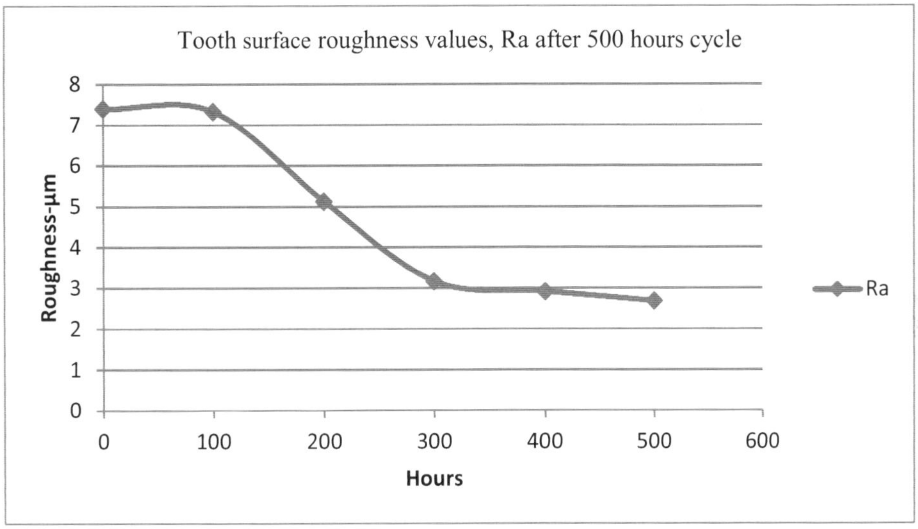

Figure 10. Surface Roughness of Marked Tooth after 500 Hours

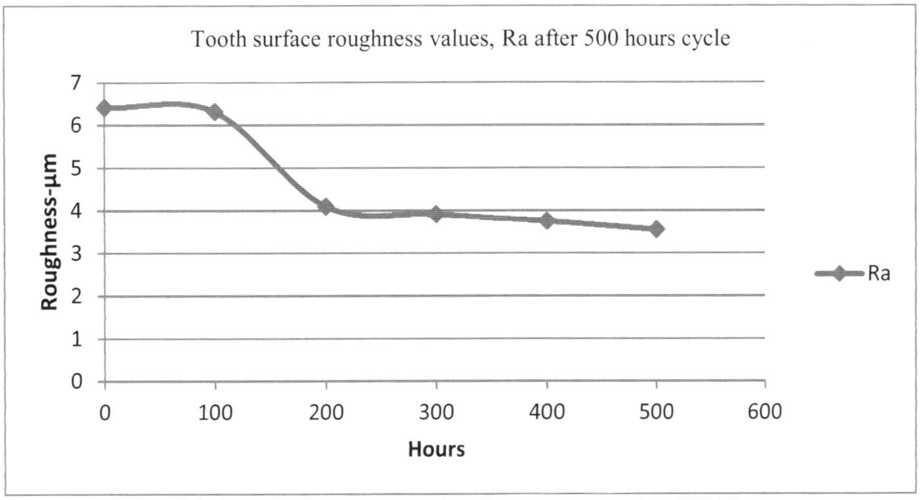

Figure 11. Surface Roughness of Adjacent Tooth after 500 Hours

IOP Conf. Series: Materials Science and Engineering **815** (2020) 012005 doi:10.1088/1757-899X/815/1/012005

The tooth surface becomes smoother after a certain number of testing cycles, as shown and explained in the surface roughness results. These results in a good agreement with research found [8].

4.3. Trend Analysis
From results, it shows that surface roughness analysis trend for marked and adjacent tooth are steadily declining after 500 hours of running meanwhile the amplitude of vibration trend are continuing increase after 500 hours of cycle.

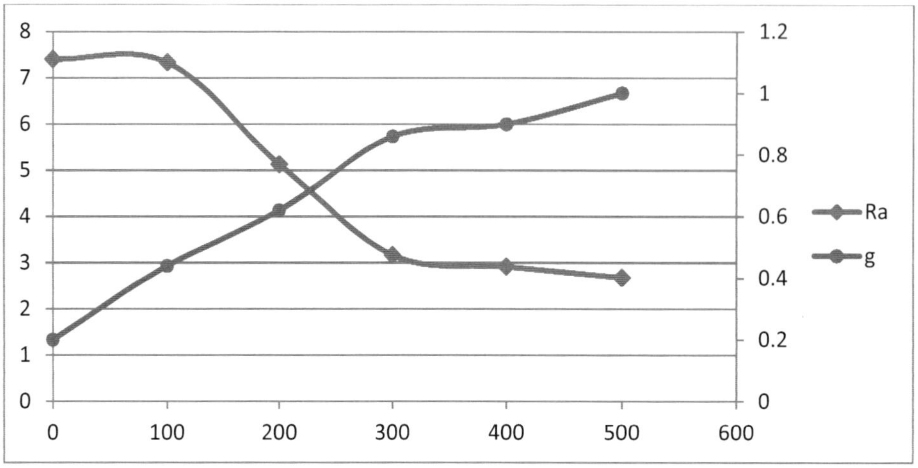

Figure 12. Trend Analysis for Marked Tooth

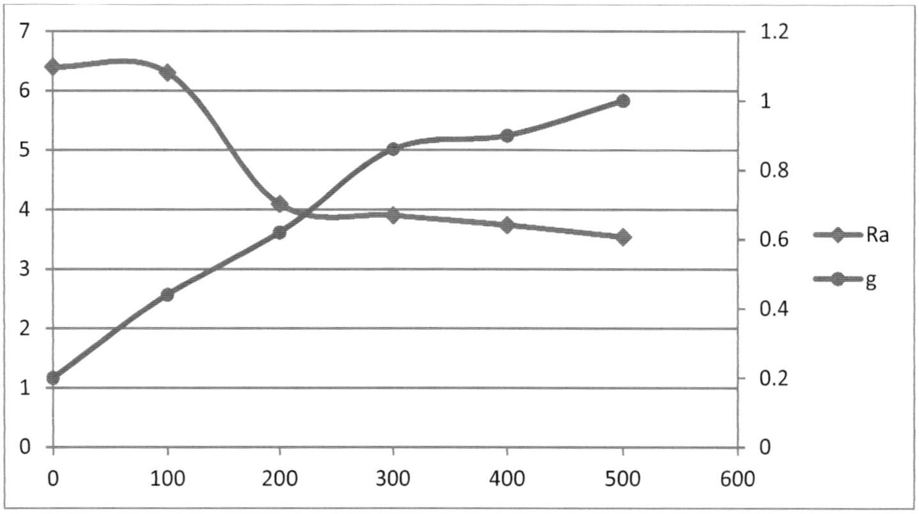

Figure 13. Trend Analysis for Extended Tooth

Figure 14 and figure 15 show the correlation between gear surface roughness and vibration amplitude after 500 hours of cycle.

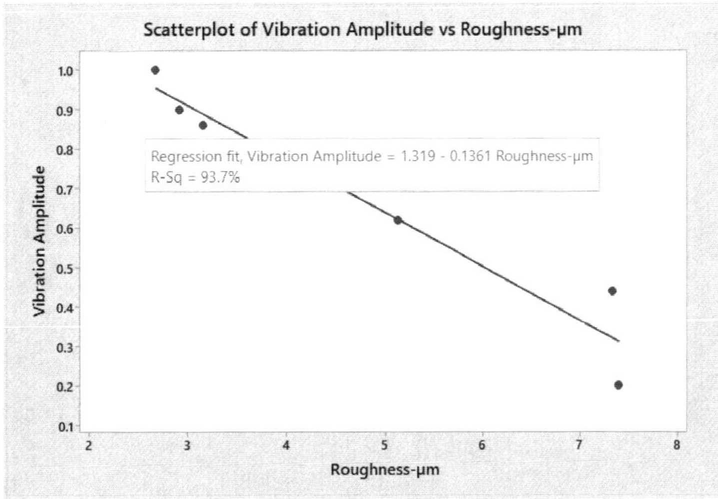

Figure 14. Correlation for Marked Tooth

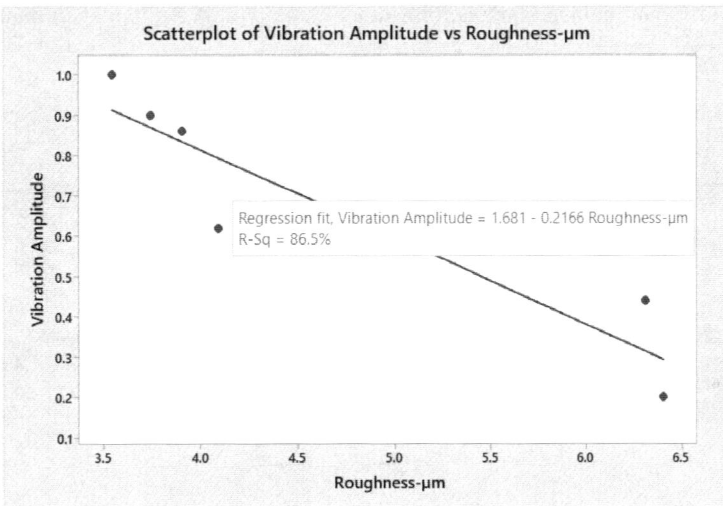

Figure 15. Correlation for Adjacent Tooth

From these results, it shows that surface roughness of tooth gear and vibration amplitude correlate with each other by 93.7% for marked tooth and 86.5% for adjacent tooth.

5. Conclusion

Based on the results obtained and presented in the previous chapters the following conclusions can be drawn. The comparison of the surface roughness measurements for both marked and adjacent tooth before and after the test runs shows that a steady trend of decrease.
The frequency of vibration data obtained through the collection of data from LMS show that the values of vibration data illustrate that amplitude trend increasing after each cycle run. For 500 hours measurement, the correlation of surface roughness of gear teeth with the amplitude of vibration is 93.7% for marked tooth

IOP Conf. Series: Materials Science and Engineering **815** (2020) 012005 doi:10.1088/1757-899X/815/1/012005

and 86.5% for adjacent tooth, this indicates the strongest possible agreement that when surface roughness of the gear teeth decreasing, the vibration amplitude is higher.

6. References

[1] Chen, J., Fang, Q., & Li, P. (2015). Effect of grinding wheel spindle vibration on surface roughness and subsurface damage in brittle material grinding. *International Journal of Machine Tools and Manufacture*, *91*, 12–23. https://doi.org/10.1016/j.ijmachtools.2015.01.003

[2] Ai-Min Y., Christophe T., & Jean-Claude G. (2004). Damage Detection in Rotating Machinery Using Statistical Methods. PCA Analysis and Autocorrelation Matrix, 11-13.

[3] Gomaa, F. R., Khader, K. M., & Eissa, M. A. (2016). Fault Diagnosis of Rotating Machinery based on vibration analysis. *International Journal of Advanced Engineering and Global Technology*, *4*(1), 1571–1586.

[4] Hocine B., Salah B., & M. S. Boucherit, (2012). Application of Wavelet Transform for Fault Diagnosis in Rotating Machinery. International Journal of Machine Learning and Computing, Vol. 2, No. 1.

[5] A.A Ibrahim, S.M.Abdel-Rahman, M. Z. Zahran, H.H.ELMongey. An Investigation for Machinery Fault Diagnosis Using Adaptive Wavelet

[6] Hassui, A., and Diniz, A. E. (2003). Correlating surface roughness and vibration on plunge cylindrical grinding of steel. Int. J. Mach. Tools Manuf., 43(8), pp.855–862.

[7] Yang, Y., Smith, W. A., Borghesani, P., Peng, Z., & Randall, R. B. (2015). Detecting changes in gear surface roughness using vibration signals. *Acoustics 2015 Hunter Valley*, 1–10.

[8] Francisco, A. R. L. (2013). Effects of Variable Load and Rotational Speed Conditions on Gear Micropitting in Wind Turbine Gearboxes. *Journal of Chemical Information and Modeling*, *53*(9), 1689–1699. https://doi.org/10.1017/CBO9781107415324.004

IOP Conf. Series: Materials Science and Engineering **815** (2020) 012006 doi:10.1088/1757-899X/815/1/012006

Frequency Domain Technique for Characterizing Spur Gears Defect Pattern

M N A Hamid[1]*, M Ismail[1], Z N M Yusuf[1]

[1]Universiti Kuala Lumpur Malaysian Spanish Institute, 09000 Kulim, Kedah, Malaysia

*Corresponding author: mnajib@unikl.edu.my

Abstract. This paper presents information of detecting the gear using frequency domain technique in vibration analysis. In this study, vibration signal is very crucial in determining signal from defective gear. The quality of the signal displayed in the frequency spectrum are depends on the number of sampling frequency taken into account during acquiring the data. Insufficient number of samples may acquire poor signal in contrast with sufficient number of samples. The objective of this research emphasizes the various frequency resolution and centroid shifting frequency location to characterize the normal and defective gear. The development of test rig with a constant speed of 1493 RPM is designed and consist of simple pinion and gear. The DAQ (NI) with algorithm block diagram developed using LabVIEW is used to acquire the frequency domain data. It is found that by varying the number of samples will affect the resolution of the signal and subsequently tune the shifting centroid of frequency. By calculating the centroid shifting frequency have indicate that the amplitude of normal gear is 0.0128g, defect gear (broken tooth) is 0.052 g and defect gear (broken tooth 90°) is 0.0938g. The amplitude of normal gear, defect gear (broken tooth) and defect gear (broken tooth 90°) are located at frequency 745.829Hz, 746.757Hz and 746.2Hz respectively.

1. Introduction

The gear is mechanical component used to transmit power from one shaft to another shaft in machinery [1]. The smooth operation and high efficiency of gears are necessary for normal running of machinery. Therefore, the contact between gears have a tendency to produce wear, fatigue, subsequently tooth failure. The gear damage detection is a main topic in the field of condition base monitoring (CBM) and fault diagnosis. A number of methods, such as time domain analysis, frequency domain analysis, time- frequency analysis and pattern recognition, have shown a high potential for the vibration signal and it can show a typical signal pattern produced by a localized fault in the various components of rotating machine. [2] This project base on condition monitoring technique of a spur gear fault detection using frequency domain technique in vibration analysis. Spur gears are the most common type of gears. They have straight teeth and are mounted on parallel shafts. Sometimes, many spur gears are used at once to create very large gear reductions. The application using spur gear are metal cutting machines, automobile gear boxes, marine engines, mechanical clocks and watches, fuel pumps, washing machines, gear motors and gear pumps, Material handling equipment's etc. The vibration signals were acquired from gearboxes and used to simulate various faults on spur gear tooth. Vibration signals were applied to monitor a normal and various fault conditions of a spur gear such as normal, defect gear (broken tooth) and defect gear (broken tooth 90°) by using LabVIEW. Signal produced by meshing gears is very difficult to characterize its pattern using vibration analysis base on condition monitoring DAQ (NI) of rotating machinery with the presence of many defects on gear teeth become more difficult to characterize its behaviour. Therefore, finding the resolution of the gear signal may assist in determining the defect behavior of the gears.

2. Literature Review

This section will discuss the knowledge gained from the article and journal reviewed. The purpose of this review is to link between the frequency domain techniques in vibration analysis with the fabricate gear test rig using spur gear for the lab uses. All findings have been summarized below.

2.1 Gear Failures

The engine defects tend to happen when a device works under conditions of increased pressure. Local faults are the more hazardous because, once launched, they tend to evolve quickly and generally have important impacts on the transmission of energy. If not identified soon, the most significant local faults can be drastic implications with teeth breakage, pitting and scoring. [3]

2.2 Vibration Signal

Vibration signals and acoustic emissions are the most prevalent waveform information in status surveillance. Ultrasonic signals, engine present, partial release are another waveform information. There are two primary types of stationary waveform information assessment in the literature which is analysis for time-domain and evaluation of frequency-domain. [1]

2.2.1 Time-Domain Analysis

Analysis of the time-domain is focused straight on the time waveform itself. Traditional time-domain assessment calculates characteristics from time waveform measurements such as mean, peak, peak-to-peak ratio, normal deviation, crest factor, and high-order data (root mean square, skew ness, kurtosis, etc.). Usually these characteristics are called time-domain characteristics. Time Synchronous Average (TSA) is a common time-domain analysis method. TSA's concept is to use the raw signal's ensemble average over a range of developments in an effort to extract or decrease noise and impacts from other sources to improve the interesting signal parts. [1]

2.2.2 Frequency-Domain Analysis

Analysis of the frequency-domain is focused on the transformed frequency domain signal. The benefit of frequency-domain assessment over time-domain assessment is its capacity to readily recognize and isolate some interesting frequency elements. The scope assessment by means of fast Fourier transform (FFT) is the most commonly used standard assessment. The primary concept of range assessment is either to look at the entire range or to look carefully at some important frequency parts and thus to remove characteristics from the signal [1]

2.3 Test Rig

It has been built to track the condition of the spur gear. This set-up consists of a single-stage gearbox powered by a switching electric motor (30 kW DC) and a controller designed to control engine speeds in the range of 0-3000 r / min. The test gearbox consists of a pair of standard involute profiled spur gears and bearings (URB32306 for the driving shaft and URB30307 for the driven shaft). Lovejoy coupling attaches the engine to the gearbox output shaft. The gearbox's output shaft connects the torque detector shaft by

Lovejoy coupling. An eddy current dynamometer (consisting of an LSG 2010 controller) coupled with the output shaft will apply torque (1-75 Nm) on the gears. [4]

2.4 LabVIEW

Lab VIEW (Laboratory Virtual Instrument Engineering Workbench) is a graphical programming language mainly used to obtain, calculate and analyse information. In this research, the development of a device for measuring and analysing vibration using Lab VIEW. The DAQ (Data Acquisition) input is in the Time domain for this programming. Using the FFT (Fast Fourier Transform) analyser program converts the time domain graph to the frequency domain and finally picks up the frequency corresponding to the highest amplitude or peak amplitude using the peak detector after the FFT analyser. [5]

3. Methodology

3.1 Flow Chart

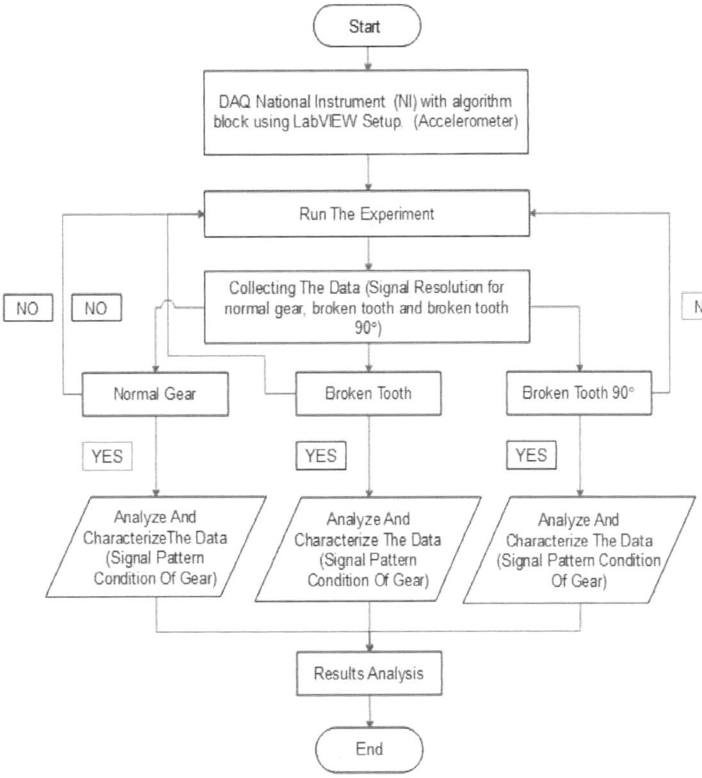

Figure 1. Flow Chart

3.2 Experimental Rig Setup

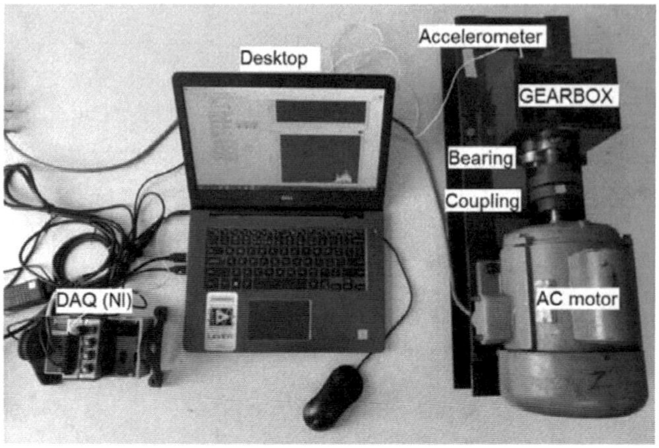

Figure 2. The experimental setup used for the gearbox testing

Figure 3. Normal gear

Figure 4. Gear Fault

The simple gear test rig is developed to investigate the condition of the spur gear through the frequency domain technique. In the gearbox test rig, the normal and specific fault conditions of the spur gear are monitored shown in figure 3 and figure 4. The accurate and reliable data were gained from the gearboxes through several signal processing technique, as shown in figure 2. The main objective of these experimental studies is to monitor the condition of the automobile gearbox running at the same speed. The gearbox was running at 1493 RPM. Shear accelerometer with sensitivity of 10.19 mV/g or 1.039 mV/m/s^2 is positioned on the housing bearing of the shaft rotate to provide the vibration signatures of the gearboxes. The gear ratio helps to determine the output and input of the speed as well as the torque and power. The gear ratio can be calculated by using equation (1). It is vital to determine the gear ratio respect to the individual spur gear rotating speed to determine gear mesh frequency in order to locate the appropriate frequency band as given in equation (2).

$$G.R = \frac{D_G}{D_P} \tag{1}$$

GR = Gear Ratio
DG = Diameter of Gear (mm)
Dp = Diameter of Pinion (mm)

$$G.M.F = \frac{T.N}{60} \tag{2}$$

GMF= Gear Mesh Frequency (Hz)
T = Number of Gear Teeth
N = Shaft Speed (rpm)

3.3 Centroid Shifting Frequency

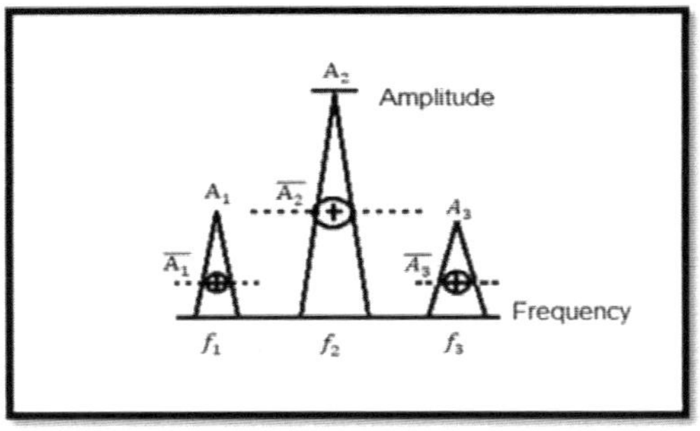

Figure 5. Signal shifting centroid

$$\bar{y} = \frac{\sum_{i=1}^{3} \bar{A}_i A_i}{\sum_{i=1}^{3} A_i} \tag{3}$$

$$\bar{F} = \frac{\sum_{i=1}^{3} f_i A_i}{\sum_{i=1}^{3} A_i} \tag{4}$$

To calculate the centroid of the defect gear as recognition feature to characterize the gear type defect. The resolution of the signal produced by the defect gear will provide specific feature that can be easily detected. By zooming at the gear mesh frequency region, the calculation of the centroid of gear defect signal can be done easily using the equation (1) and (2). Therefore, for each type of defect produce light significant of signal feature that can be differentiated its characteristic.

3.4 Resolution Setup

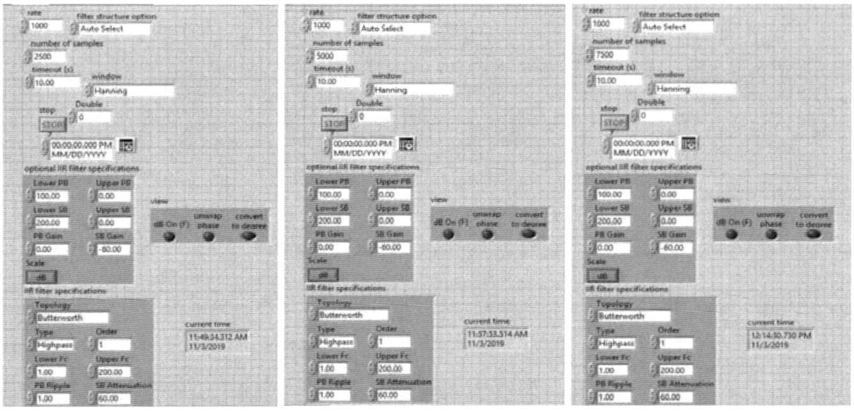

Figure 6. Resolution setup using LabVIEW

Figure 6 show the resolution setup using LabVIEW by varying the number of sample from 2500, 5000 and 7500. To get a sufficient of data, the technique use was varying the number of sample. If the data is insufficient, the data resolution is ambiguous. The bandwidth frequency is fixed at 1000 Hz for both gear; normal and fault gear. The resolution setup using LabVIEW is to emphasize the various frequency resolution and centroid shifting frequency to characterize the defective gear.

4. Results and discussion

In this section the standard gear outputs are calculated as no defect condition. To get the quality of the signal displayed in the frequency spectrum, three different number of samples was taken from 2500, 5000 and 7500 using DAQ (NI) in LabVIEW. The detail of graph illustration is tabulated below:

Table 1. Sample LabVIEW normal gear table

No. of gear	Rate	Number of samples	Gear condition	G.M.F centroid	Result $\bar{y}$ =	$\overline{F}$ =
1	1000	2500	Normal gear	A_1 = 0.0351 A_2 = 0.0360 A_3 = 0.0356	0.0178g	744.440 Hz
2	1000	5000	Normal gear	A_1 = 0.0292 A_2 = 0.0193 A_3 = 0.0427	0.0168g	745.526 Hz
3	1000	7500	Normal gear	A_1 = 0.0252 A_2 = 0.0310 A_3 = 0.00970	0.0128g	745.829 Hz

4.1 Normal gear

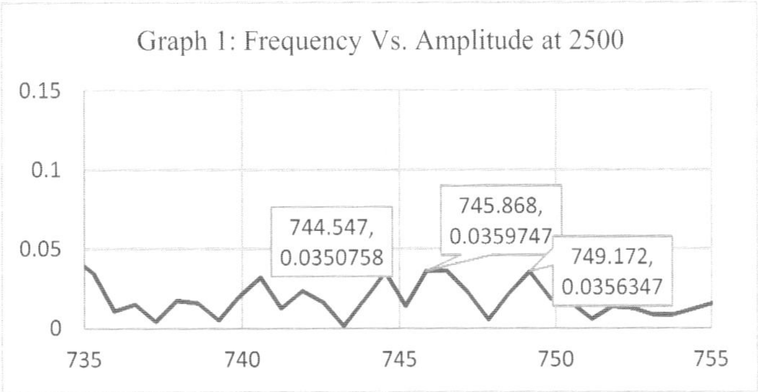

Figure 7. Frequency domain vibration signal of the normal gear at 2500 number of samples

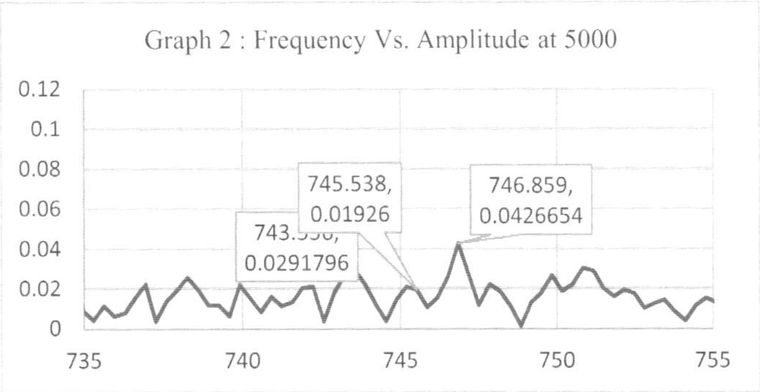

Figure 8. Frequency domain vibration signal of the normal gear at 5000 number of samples

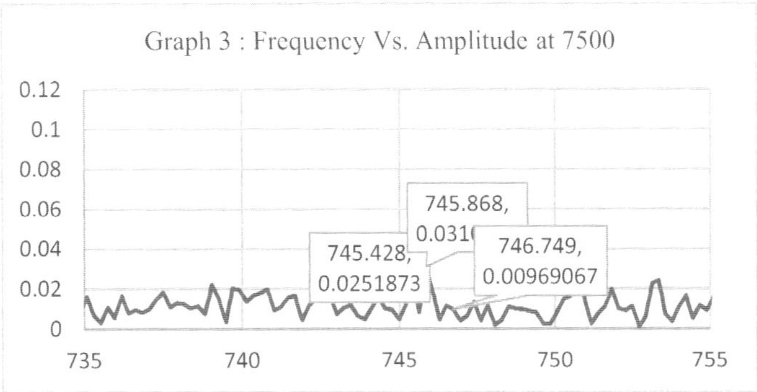

Figure 9. Frequency domain vibration signal of the normal gear at 7500 number of samples

The finest resolution of normal gear frequency is 745.829 Hz because the value is almost precise with gear mesh frequency that has been calculated. Other than mesh frequency, the resolution is also determined by the smallest value of side band from the graph. From this data, the frequency domain technique for characterizing spur gears defect pattern can be detected.

4.2 Broken tooth

In this section the default gear outputs are calculated to define the gear in a defect condition (broken tooth). To get the quality of the signal displayed in the frequency spectrum, three different number of samples was taken from 2500, 5000 and 7500 using DAQ (NI) in LabVIEW. The detail of graph illustration is tabulated below:

Table 2. Sample Lab View defect gear (Broken Tooth) table

No. of gear	Rate	Number of samples	Gear condition	G.M.F centroid	Result $\bar{y} =$	$\bar{F} =$
1	1000	2500	Broken tooth	$A_1 = 0.0527$ $A_2 = 0.166$ $A_3 = 0.0364$	0.0619 g	746.360 Hz
2	1000	5000	Broken tooth	$A_1 = 0.0477$ $A_2 = 0.134$ $A_3 = 0.050$	0.0491g	744.526 Hz
3	1000	7500	Broken tooth	$A_1 = 0.0525$ $A_2 = 0.0526$ $A_3 = 0.143$	0.0524 g	746.757Hz

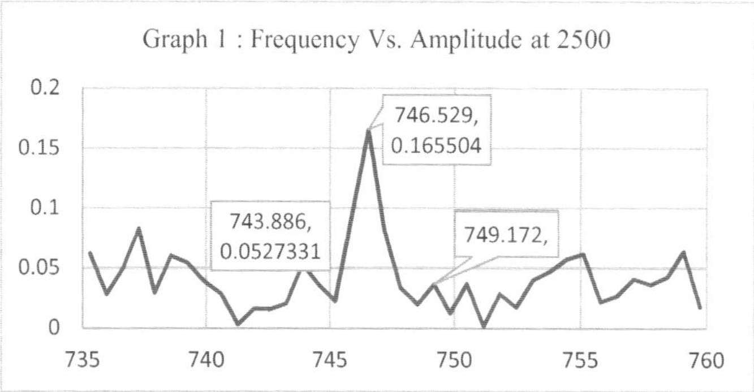

Figure 10. Frequency domain vibration signal of the broken tooth at 2500 number of samples

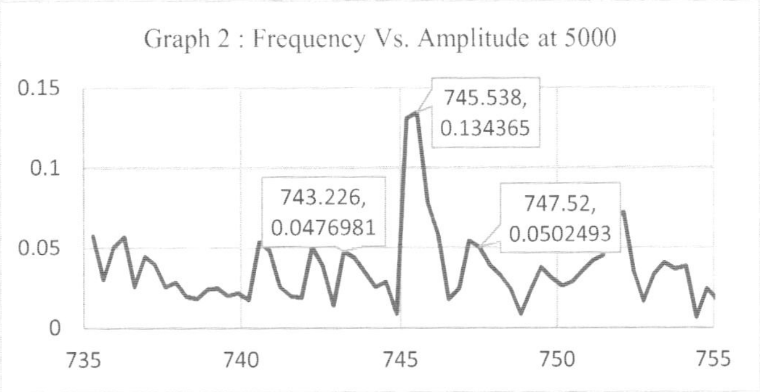

Figure 11. Frequency domain vibration signal of the broken tooth at 5000 number of samples

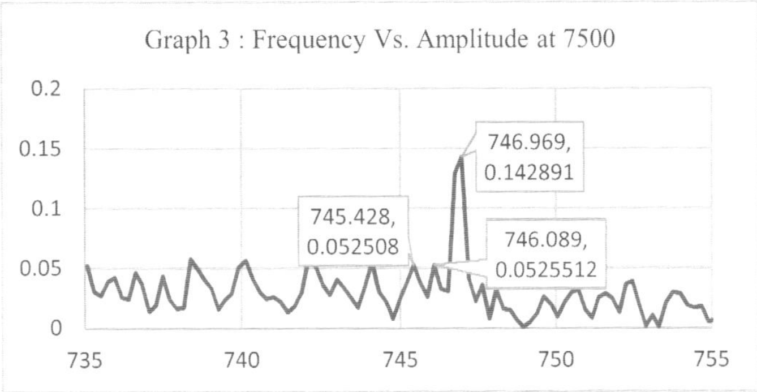

Figure 12. Frequency domain vibration signal of the broken tooth at 7500 number of samples

The finest resolution of fault gear frequency is 746.757Hz because the value is approximately with gear mesh frequency that has been calculated. Other than mesh frequency, the resolution is also determined by the smallest value of side band from the graph. From this data, the frequency domain technique for characterizing spur gears defect pattern can be detected.

4.3 Broken tooth 90°

In this section the default gear outputs are calculated to define the gear in a defect condition (broken tooth 90°). To get the quality of the signal displayed in the frequency spectrum, three different number of samples was taken from 2500, 5000 and 7500 using DAQ (NI) in LabVIEW. The detail of graph illustration is tabulated below:

Table 3. Sample Lab View defect gear (Broken Tooth 90°) table

No. of gear	Rate	Number of samples	Gear condition	G.M.F Centroid	Result $\bar{y}=$	$\overline{F}=$
1	1000	2500	Broken tooth 90°	$A_1=0.0804$ $A_2=0.224$ $A_3=0.0423$	0.0843 g	745.191Hz
2	1000	5000	Broken tooth 90°	$A_1=0.0514$ $A_2=0.228$ $A_3=0.0437$	0.0875g	745.213Hz
3	1000	7500	Broken tooth 90°	$A_1=0.026$ $A_2=0.234$ $A_3=0.0475$	0.0938g	746.2Hz

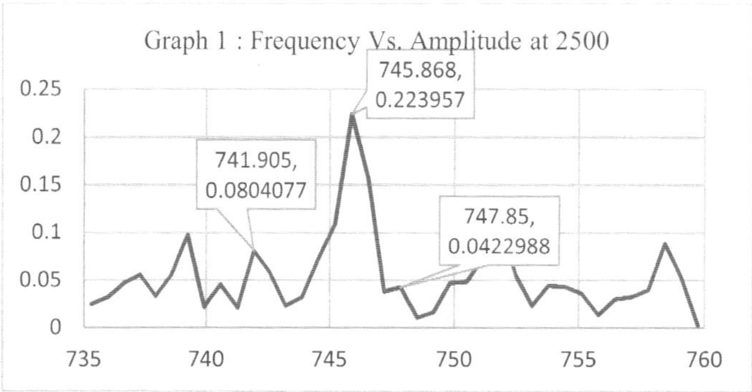

Figure 13. Frequency domain vibration signal of the broken tooth 90° at 2500 number of samples

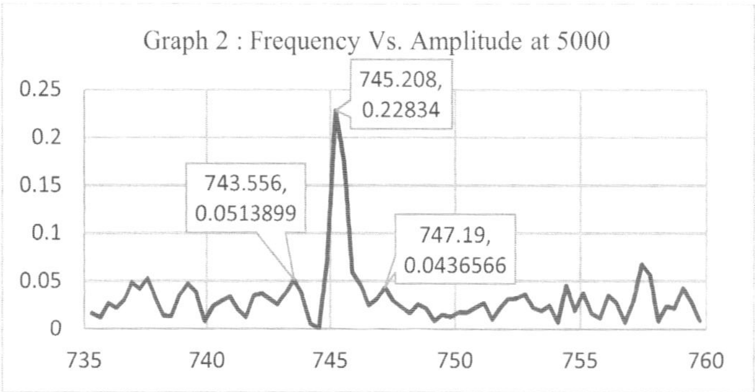

Figure 14. Frequency domain vibration signal of the broken tooth 90° at 5000 number of samples

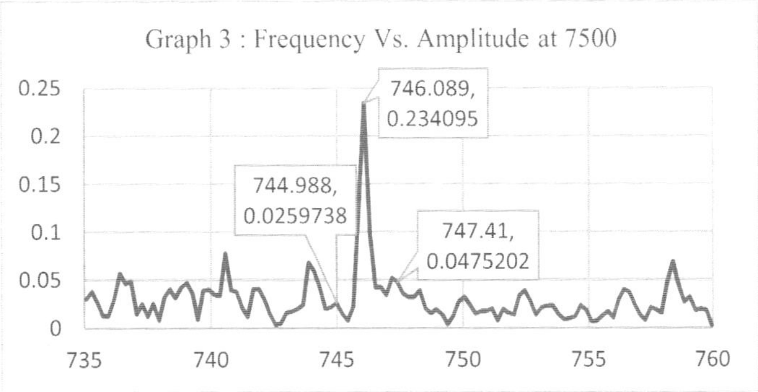

Figure 15. Frequency domain vibration signal of the broken tooth 90° at 7500 number of samples

The finest resolution of fault gear frequency is 746.2Hz because the value is approximately with gear mesh frequency that has been calculated. Other than mesh frequency, the resolution is also determined by the smallest value of side band from the graph. From this data, the frequency domain technique for characterizing spur gears defect pattern can be detected.

5. Conclusion

This paper, through a range of experimental results, showed that the technique of vibration signal analysis has the potential to investigate the behaviour of the gearbox rotation. By using LabVIEW with National Instrument (NI) DAQ, the vibration signal of a normal gear and default gear has produced a best signal resolution for gear mesh of frequency domain technique. The signal pattern from the result obtained can be used to analyse the condition of gear. Show that the result for shifting frequency based on calculation for normal gear is 745.829 Hz. As for gear fault (broken tooth), the value of shifting frequency is 744.562 Hz meanwhile the value for gear fault (broken tooth 90°) is 746.2Hz. Compared to normal gear, it can be seen that the high signal shifting frequency of fault gear is increased. All parameters have higher values in tooth breakage gear than normal gear [6].

6. Reference

[1] El-morsy, M. S., Abouel-seoud, S. A., & Rabeih, E.-adl. (2010). Geared System Condition Diagnostics Via Torsional Vibration Measurement.
[2] Klinchaeam, S., Ajavakom, N., & Yongchareon, W. (2014). Fault detection of a spur gear using vibration signal with multivariable statistical parameters.
[3] Al-arbi, S. K. (2012). Condition monitoring of gear systems using vibration analysis Salem K. Al-Arbi A thesis submitted to the University of Huddersfield in partial fulfilment of the requirements for the degree of University of Huddersfield.
[4] Shah, H., & Hirani, H. (2014). Online condition monitoring of spur gears. doi: 10.1784/204764214813883298
[5] Desai, A. K., Bharate, A. G., Rane, V. P., Gholap, S. V., & Mitra, A. C. (2017). Lab VIEW Programming for Vibration Analysis. doi: 10.9790/1684-17010010105
[6] Alshammari, S. A. M., Makrahy, M. M., & Ghazaly, N. M. (2019). Fault Diagnosis of Helical Gear through Various Vibration Techniques in Automotive Gearbox. doi: 10.12691/jmdv-7-1-3

IOP Conf. Series: Materials Science and Engineering **815** (2020) 012007 doi:10.1088/1757-899X/815/1/012007

Numerical Analysis on Vibro-acoustic Characterisation of Hybrid Honeycomb Structure with Flow Permeability

Zhen Wei Tee and Choe Yung Teoh*

Faculty of Engineering and Technology, Tunku Abdul Rahman University College, Kuala Lumpur, Malaysia

*Corresponding author: teohcy@tarc.edu.my

Abstract. The application of conventional noise attenuating structures was often limited when the exchange of air was required due to the trade-off between their noise attenuation properties and ventilation efficiency. The large physical footprint of conventional noise attenuating structures with flow permeability had also limited their applications. This paper investigated the use of corrugation structure in lightweight honeycomb sandwich panels with flow permeability to improve noise attenuation while preserving ventilation. The vibro-acoustic property of the honeycomb structure with flow permeability at low to medium frequency range (50-3000 Hz) was investigated through numerical simulation. Two different hybrid panel designs with U-shaped and V-shaped corrugations in the honeycomb core had been proposed and were found to exhibit superior sound transmission loss (STL) compared to the honeycomb panel without significantly affecting the flow performance. It was found that in the frequency range of 50-3000 Hz, the unweighted STL of the U-shaped corrugation honeycomb hybrid panel (UCHP) and V-shaped corrugation honeycomb hybrid panel (VCHP) was 19 dB and 25 dB higher than the honeycomb panel respectively with negligible effect on the flow performance. The proposed hybrid panel showed promising noise attenuation applications which involved fluid flows such as in fan and engine noise reduction.

1. Introduction

Noise control measures such as using sound isolators, sound absorbers, mufflers, baffles, as well as redesigning of sound sources were often used to reduce the excessive noise as well as their adverse impact on the human and environment. These conventional noise attenuation material and structures such as high-density panels and micro-perforated panel (MPP) with rigid backing relied on sound reflection or sound absorption to prevent the transmission of sound across a boundary. However, these sound attenuation materials or structures often stopped both the acoustic wave transmission and steady fluid flow across the boundary which eliminated the passage of air. As a result, this limited their applications when the exchange of air was necessary or advantageous such as in noise control of cooling fans where free circulation of air was important in order to allow heat transfer and dissipation. Besides, noise reduction in the low frequency region (<500 Hz) using conventional approaches was challenging and often involved increasing the thickness of acoustic barrier layer as the transmission of noise through a material was inversely proportional to the product of the frequency, material density and material thickness [1].

To date, several approaches and acoustic enclosure with openings had been proposed or designed to block sound while maintaining the transport of air. For instance, diffraction type resonator with negative effective bulk modulus was utilized in designing a soundproof window that was transparent to airflow [2]. The use of an acoustic metamaterial with a large orifice to allow airflow while achieving narrow-band sound insulation through the interaction of resonating field of four membranes had also been demonstrated [3]. Several studies had also applied active noise control to simultaneously achieve acoustic insulation and natural ventilation [4, 5]. Although the transmission loss or sound absorption properties of the mentioned designs were high, the structures generally had a large physical footprint or require the use of complex software and hardware which limited their application for real-world problems such as reducing noise from fans and compressors [6].

Among various lightweight structure, honeycomb panel was one of the most commonly used structure in the industries with wide application due to its high strength to weight ratio. Although the characterisation of the honeycomb structure had been subjected to many scientific works from the mechanical viewpoint, research on the acoustic characteristics of honeycomb structure was still lacking, especially in the area involving the transfer of fluid. Several studies had also demonstrated that the honeycomb structure had promising potential in achieving noise attenuation while allowing for the transfer of air [6, 7]. Besides, hybrid-cored sandwich panels had also been shown to possess superior mechanical strength and impact energy absorption ability which could improve the stiffness of the panel as well as its acoustic properties [8, 9].

This paper investigated the use of hybrid honeycomb structure to achieve noise attenuation while maintaining air permeability. The vibro-acoustic properties of the honeycomb panel at low to medium frequency range (50-3000 Hz), as well as its flow performance, was studied and analysed through numerical simulations with commercial ANSYS software. The improvement on sound attenuation properties via the use of corrugation structure to form the hybrid honeycomb structure was then investigated.

2. Methodology

Numerical analysis with finite element method (FEM) was conducted to investigate the vibro-acoustic performance of the honeycomb panel design with flow permeability. Numerical method was used as it allowed the study of complex structure or situation that would otherwise be too time-consuming or difficult to be solved analytically. Besides, with finite element method, the bi-directional coupling of the fluid-structure interaction between a structural body and a fluid such as air could be taken into account in the acoustic analysis.

2.1. Honeycomb panel design

A honeycomb panel (HP) which consisted of two faceplates with the honeycomb cores in between as shown in figure 1, was developed and used as the baseline for the numerical analysis. The dimension of the honeycomb panel was summarized in table 1 and the geometric structure of the honeycomb cell was illustrated in figure 2. To allow for air permeability, both the top and bottom facings of the honeycomb panel were perforated with macro-sized holes of 10 mm. The perforated holes were located concentrically with the honeycomb core so that air would be allowed to flow through the honeycomb cavity before passing through the panel via the perforated hole at the bottom faceplate as shown in figure 3. The open area ratio of the honeycomb panel developed was also computed as 14.1% using equation (1):

$$\text{OAR} = \frac{A_{open}}{A_{total}} \text{ x } 100\% \qquad (1)$$

where A_{open} and A_{total} was the open area and total area of the honeycomb panel respectively.

Acrylonitrile butadiene styrene (ABS) plastic was selected as the material for the honeycomb panel to be used in the numerical analysis due to its low density. Besides, as ABS was a common material in 3D printing, the honeycomb panel could be fabricated easily through 3D printing for experimental validation of the simulation results in the future study.

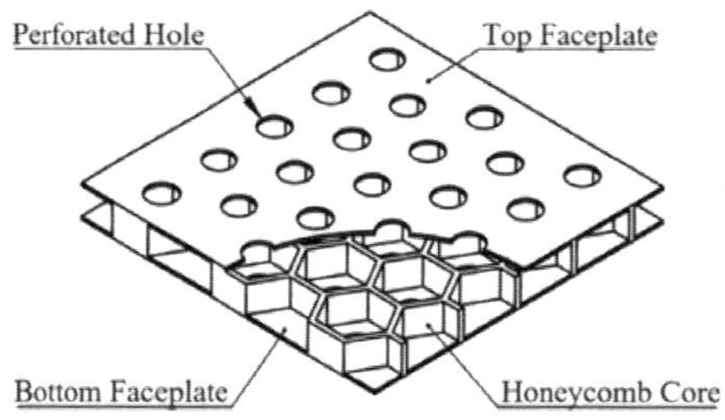

Figure 1. Schematic diagram of honeycomb sandwich panel with flow permeability

Table 1. Dimension of honeycomb panel

Dimension	Value
Faceplate Length, L_f	114.32 mm
Faceplate Width, W_f	112.00 mm
Faceplate Thickness, t_f	1.00 mm
Diameter of perforated hole, d	10.00 mm
Honeycomb cell size, H	20.00 mm
Thickness of honeycomb cell wall, b	2.00 mm
Thickness of honeycomb core, t_c	10.00 mm
Internal angle of honeycomb cell, θ	60°

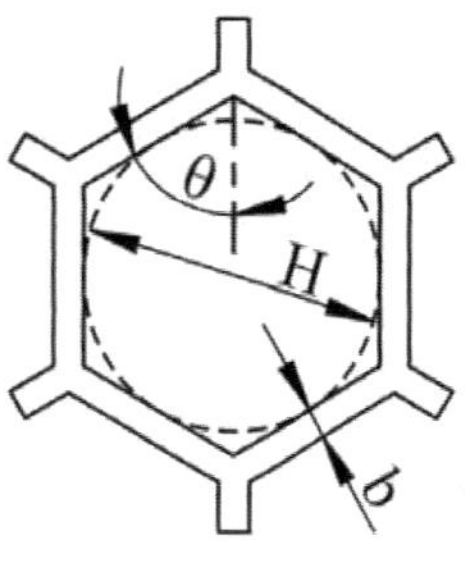

Figure 2. Geometric structure of honeycomb cell.

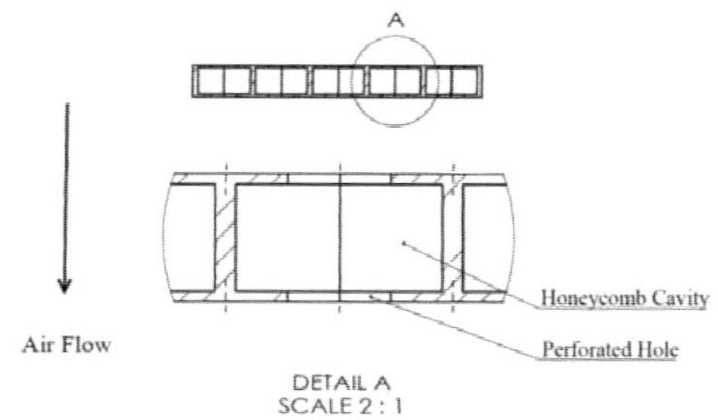

Figure 3. Detail view of perforated hole and honeycomb cavity.

2.2. Vibro-acoustic simulation model

The vibro-acoustic numerical simulation was conducted using ANSYS Harmonica Acoustic solver to calculate the acoustic response of a system due to excitation by a sinusoidal varying acoustic pressure where the excitation was continuous at a constant frequency. The acoustic simulation was conducted over a frequency range of 50-3000 Hz in order to determine the sound transmission loss (STL) of the honeycomb panel at low to medium frequency range. The fluid and material properties used for the numerical simulation were also summarised in table 2.

The simulation model consisted of an infinite duct that was divided by the honeycomb panel with perforated facing as shown in figure 4. Numerical analysis of a panel in a duct was used in this study

Table 2. Properties of fluid and material used for numerical analysis

Description	Symbol	Value	Unit
Fluid: Air			
Speed of sound	c_0	346.25	m/s
Density	ρ_0	1.225	kg/m^3
Viscosity	μ_0	1.7894×10^{-5}	kg/ms
Material: Acrylonitrile butadiene styrene (ABS)			
Density	ρ	1040	kg/m^3
Young Modulus	E	2.390	GPa
Poisson Ratio	v	0.399	
Bulk Modulus	K	3.944	GPa
Shear Modulus	G	0.854	GPa
Tensile Yield Strength	σ_y	41.4	MPa
Tensile Ultimate Yield Strength	σ_u	44.3	MPa
Percent Elongation	ε_e	24.3	%
Tangent Modulus	K	68.3	MPa

as it could provide physical insight for the design of the STL of a panel system and was more manageable than the partitions between two large reverberation chambers [10]. The boundary conditions applied to the numerical model for acoustic simulation were also illustrated in figure 5.

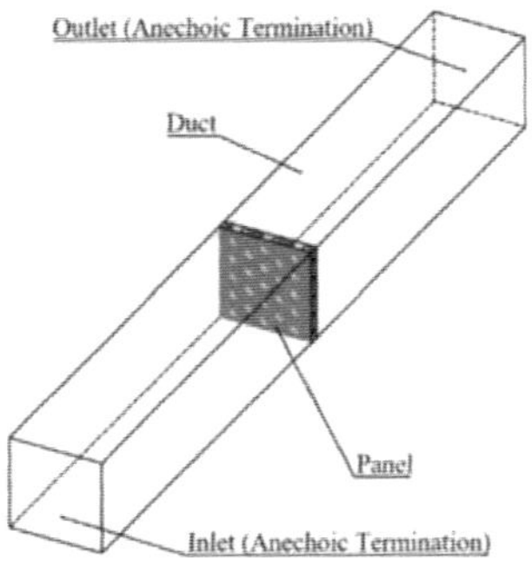

Figure 4. Simulation model for acoustic analysis.

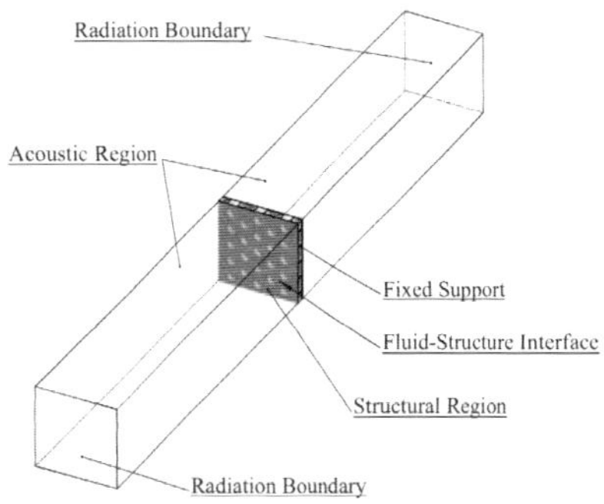

Figure 5. Boundary condition applied on numerical model.

As STL was source independent, a mass source, $\dot{M}$ with an arbitrary value of $1\ kg/m^2s$ was applied on the upstream end face of the duct (inlet) as the acoustic excitation to generate the sound wave that would propagate along the duct until it stroke the honeycomb panel [11]. The sound would be transmitted through the honeycomb panel and propagated toward the downstream end face of the duct (outlet) where it was absorbed by the anechoic termination. With the constant mass source, $\dot{M}$ of $1\ kg/m^2s$, the inlet sound pressure level, SPL_{in} was computed as 135.74 dB using equation (2):

$$SPL_{in} = 20\log_{10}\left(\frac{p}{p_0}\right) - 20\log_{10}(\sqrt{2}) \tag{2}$$

where p_0 was the reference pressure of 20 μPa and p was the equivalent incident pressure for the mass source excitation which could be expressed as:

$$p = \frac{1}{2} Q p_0 c_0 \tag{3}$$

where Q was the acoustic source volume flow rate and c_0 was the speed of sound in air. The equivalent incident acoustic pressure, p and the sound pressure level, $\mathrm{SPL_{in}}$ were computed instead of measuring the average sound pressure level at the inlet face of the simulation model as the impedance change caused by the panel would cause the acoustic energy to be reflected back upstream toward the inlet. The reflection of acoustic energy would cause the sound pressure level measured at the inlet face to increase, hence resulting in an artificially higher STL if the measured incident acoustic sound pressure of the simulation model was used [12].

The outlet sound pressure level, $\mathrm{SPL_{out}}$ of the simulation was measured to compute the STL of the honeycomb panel which could be expressed by equation (5):

$$\mathrm{STL} = 20\log_{10}\left(\frac{p_{\mathrm{incident}}}{p_{\mathrm{transmitted}}}\right) + 20\log_{10}\left(\frac{S_{\mathrm{inlet}}}{S_{\mathrm{outlet}}}\right) \tag{4}$$

$$\mathrm{STL} = SPL_{\mathrm{in}} - SPL_{\mathrm{out}} + 20\log_{10}\left(\frac{S_{\mathrm{inlet}}}{S_{\mathrm{outlet}}}\right) \tag{5}$$

where $p_{\mathrm{transmitted}}$ was the average pressure at the outlet face, S_{inlet} and S_{outlet} were the areas of the inlet and outlet face respectively. As the inlet and outlet faces of the duct were equal ($S_{\mathrm{inlet}} = S_{\mathrm{outlet}}$), equation (5) could be simplified as:

$$\mathrm{STL} = SPL_{\mathrm{in}} - SPL_{\mathrm{out}} \tag{6}$$

It should also be highlighted that plane wave conditions were assumed to exist at the inlet and outlet faces for equations (5) and (6) to be valid. The calculation of STL for non-plane wave condition would involve the use of real and imaginary part of the acoustic pressure and particle velocity, as well as the area at each node and at each frequency of the inlet and outlet faces to compute the acoustic intensity and effective nodal area in order to determine the sound power, and hence was not covered in this study [12].

The natural frequencies and mode shapes of the honeycomb panel with all lateral faces fixed were also extracted using ANSYS Modal solver. This was accomplished by solving the acoustic simulation model without the acoustic region and any excitation. The natural frequencies and mode shapes were extracted to investigate the acoustic performance of the panel at resonance.

2.3. Flow simulation model

As no mean flow could be included in the acoustic analysis due to a different formulation of wave equation was required if the mean flow was included, computational flow dynamics (CFD) simulation was conducted to analyse the flow performance of the honeycomb panel using ANSYS CFX solver. To simulate airflow in the duct, the acoustic excitation was replaced with airflow so that air would flow into the duct through the inlet and exit through the outlet after passing through the perforated facings and honeycomb cavity. The simulation was conducted at different inlet velocity, u ranging from $0.5\ \mathrm{m/s}$ to $3.0\ \mathrm{m/s}$ with a constant 5% turbulence to investigate the pressure loss at different inlet velocity. The pressure loss, ΔP due to the honeycomb panel was computed using equation (7):

$$\Delta P = P_{\mathrm{inlet}} - P_{\mathrm{outlet}} \tag{7}$$

where P_{inlet} and P_{outlet} was the inlet and outlet pressure respectively.

2.4. Hybrid honeycomb panel design

Two different hybrid honeycomb panel designs were proposed to improve the noise attenuation properties of the honeycomb panel without significantly impacting the flow performance. The improvements made were focus on the honeycomb core as the geometry and the dimension of the system would affect the sound insulation at low frequency [13, 14]. The improvements involved implementing corrugation structure in the honeycomb core to form a honeycomb-corrugation hybrid core. Two different corrugation structures which were U-shaped corrugation and V-shaped corrugation were proposed and investigated. Similar to the honeycomb panel design used in this study the corrugation structure was perforated such that it did not obstruct the airflow. In order to investigate the effect of the different corrugation structure in the honeycomb panel on the flow performance, the OAR of the two proposed panels was kept constant at 14.1%. The proposed U-shaped corrugation honeycomb hybrid panel (UCHP) and V-shaped corrugation honeycomb hybrid panel (VCHP) were illustrated in figure 6 and figure 7 respectively.

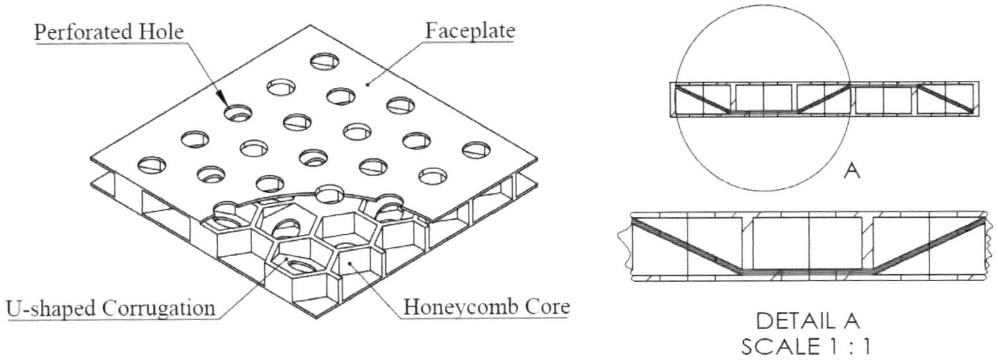

Figure 6. Schematic diagram of U-shaped corrugation honeycomb hybrid panel (UCHP)

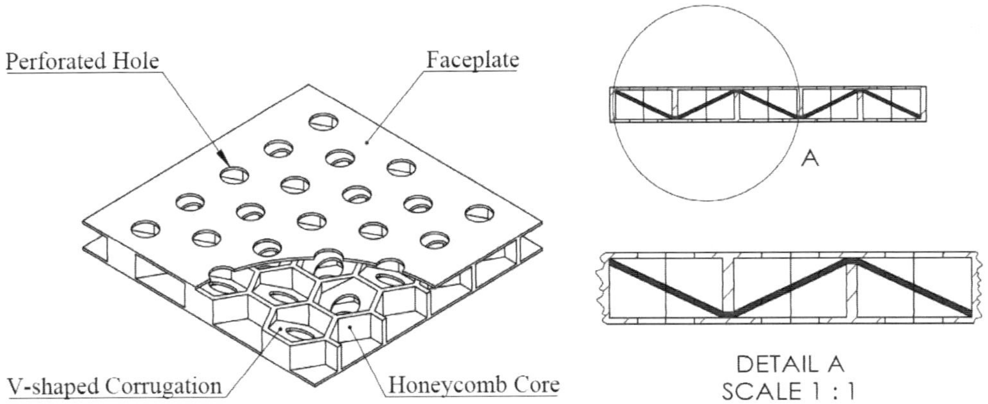

Figure 7. Schematic diagram of V-shaped corrugation honeycomb hybrid panel (VCHP)

3. Result and discussion

3.1. Modal analysis

From the modal analysis, the first six natural frequencies of the honeycomb panel (HP), UCHP and VCHP were computed and summarized in table 3. It could be observed that only the first natural frequency of the panels was located in the frequency range of 50-3000 Hz used in the acoustic simulation. This indicated that all three panels would undergo the first resonance in the acoustic simulation as their first natural frequency was located in the frequency range used in the acoustic simulation. Besides, it could be seen that at the same mode, the honeycomb panel had the highest natural frequency followed by UCHP and then VCHP, but the natural frequencies of the different panel designs were relatively close.

The mode shapes for the first natural frequency of the different panel designs were also illustrated in figure 8 and it could be seen that the UCHP and VCHP exhibited similar mode shapes with the honeycomb panel due to similar overall structure of the panels. It could also be observed that each of the panel exhibit flexural vibrations as the two faceplates bent in the same direction. Dilatational vibration which occurred when the faceplates bent in opposite direction and would result in undesirable coincidence response in the sound transmission spectrum of lightweight sandwich panels was not observed in the first six natural frequencies [15, 16].

Table 3. Natural frequencies of different panel designs

	Frequency (Hz)		
Mode	HP	UCHP	VCHP
1	1976.9	1901.2	1895.0
2	3342.6	3208.6	3186.0
3	3375.4	3344.4	3291.6
4	4449.4	4345.5	4306.0
5	4995.7	4818.0	4767.5
6	5063.4	4943.0	4944.1

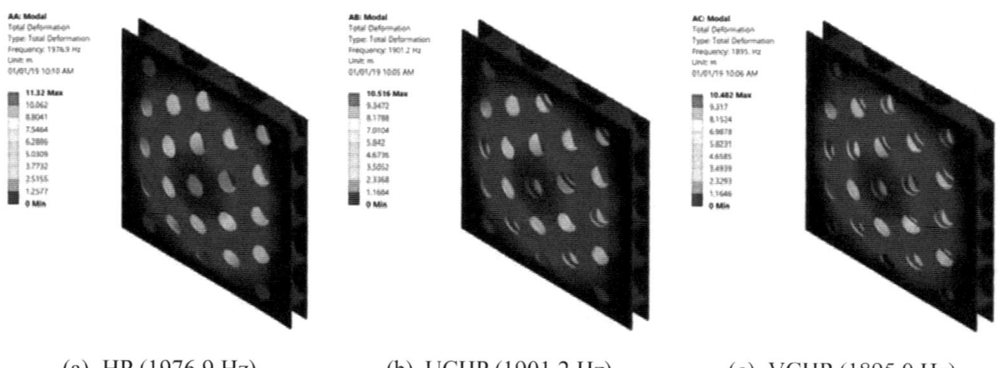

(a) HP (1976.9 Hz) (b) UCHP (1901.2 Hz) (c) VCHP (1895.0 Hz)

Figure 8. Mode shape of different panel designs at first natural frequency.

3.2. Vibro-acoustic performance

The average outlet sound pressure level, SPL_{out} was shown in figure 9 and it could be seen that the SPL_{out} for each panel design was always lower than the computed inlet sound pressure level, SPL_{in} of 135.74 dB, implying that the sound level was reduced after passing through the panel. The STL of different panel designs was also shown in figure 10 and it could be seen that in the low to medium frequency region, the STL decreased as the frequency increased. It was also observed that the STL of the VCHP was the highest, followed by the UCHP and the honeycomb panel. A rapid reduction in the STL was also observed from figure 10 near the first resonance frequency of the panel as the panel resonated with the acoustic excitation which caused a rapid reduction in sound insulation. A higher STL observed in the stiffness-controlled region which was located at the low frequency region before the first resonance frequency for UCHP and VCHP generally implied that the UCHP and VCHP had a higher stiffness compared to the honeycomb panel due to the presence of the corrugation structure which stiffness the panel and improved the STL [8, 9]. Although increasing the stiffness of the panel would generally lower the coincidence frequency in the coincidence region but it was not considered in the present study as the coincidence region was located at the higher frequency region and high frequency noise could be damped more easily using conventional noise attenuation method [17].

As the STL of the different panels varied with similar trends when the frequency increased, the unweighted average STL value for the frequency range of 50-3000 Hz was computed and tabulated in table 4. The unweighted average STL value was used to provide a single number rating of the transmission loss to evaluate the acoustic performance of the panel. It could be seen that the VCHP had the highest unweighted average SSTL, followed by UCHP and the honeycomb panel. From table 4, it could be seen that the U-shaped and V-shaped corrugation in the honeycomb core improved the unweighted average STL by approximately 19 dB and 25 dB respectively, showing a significant improvement in the noise attenuation properties.

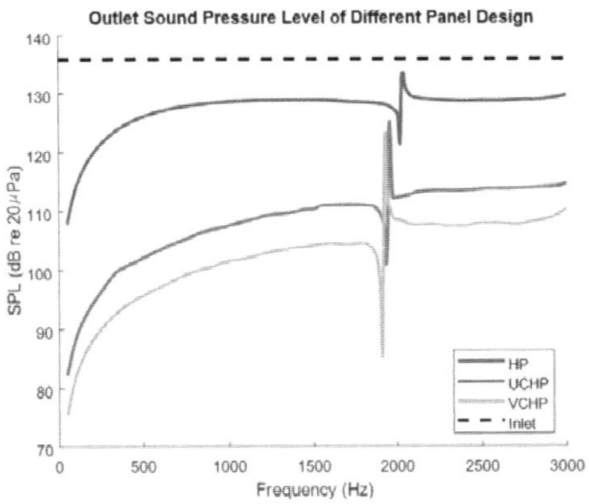

Figure 9. SPL_{out} of different panel design

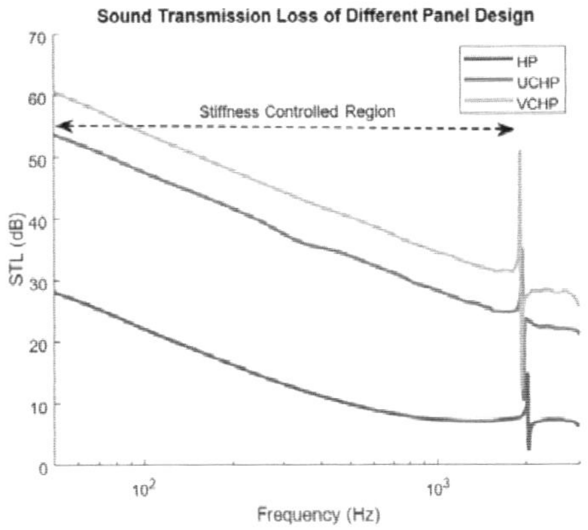

Figure 10. STL of different panel design

Table 4. Unweighted average STL (50-3000 Hz)

Panel	Unweighted Average STL (dB)
HP	8.54
UCHP	27.53
VCHP	33.58

3.3. *Flow performance*

Figure 11 illustrated the pressure distribution through the duct with different panel designs with an inlet flow velocity, u of $3\ \mathrm{m/s}$ and it could be seen that the pressure decreased when the flow approached the panel which was located from duct length, $L = 0.006\ \mathrm{m}$ to $L = -0.006\ \mathrm{m}$ as the air accelerated when passing through the perforated facings. The increase in fluid velocity when passing through the panel resulted in low pressure in the panel due to the conversion of energy as the fluid potential energy was converted into kinetic energy, The pressure was recovered at the downstream of the duct as the fluid decelerated and the excess kinetic energy was converted back into the pressure, causing the pressure to increase. From figure 11, more number of dips in pressure was also observed in UCHP and VCHP compared to the honeycomb panel as the flow experienced more contraction and expansion when flowing through the panels due to the presence of corrugated structure in the honeycomb core.

The pressure loss of different panel designs as a function of the inlet flow velocity, u was also shown in figure 12 and it could be observed that the pressure loss increased with inlet flow velocity for all the panel. It was also observed that the VCHP panel experienced the highest pressure loss, followed by the UCHP and the honeycomb panel. However, the difference between the pressure losses for the different panel design was relatively small and negligible, especially at low inlet velocity, when the open area was kept constant for all the panels. Hence, it could be concluded that the V-shaped and U-shaped corrugation in the honeycomb did not affect the flow performance of the honeycomb panel provided that the open area of the panel was kept constant.

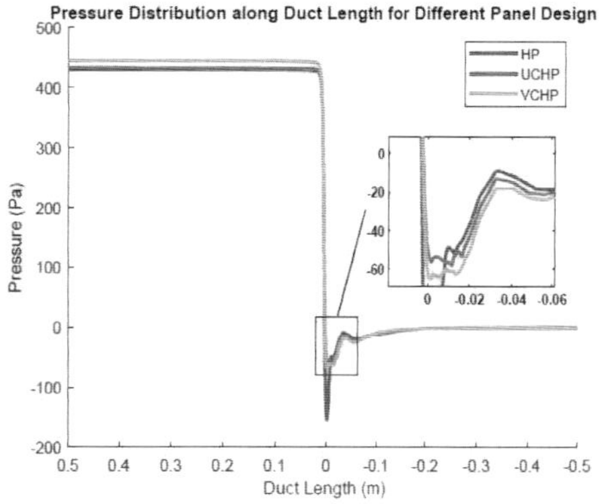

Figure 11. Pressure distribution along the duct at $u = 3.0\ \mathrm{m/s}$ for different panel design.

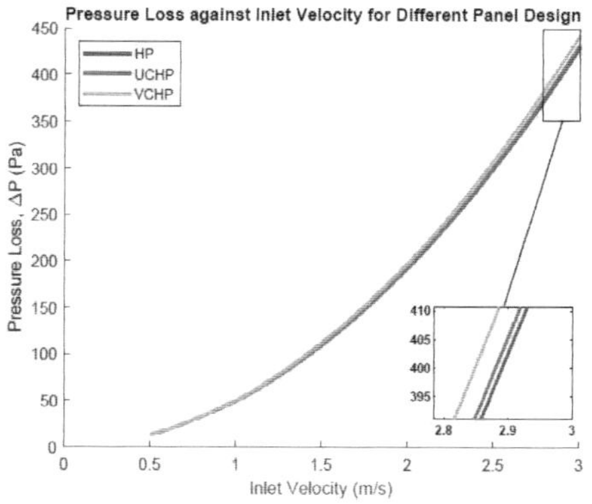

Figure 12. Pressure drop as a function of inlet velocity for different panel design.

4. Conclusion

A honeycomb sandwich panel with perforated facings was designed to allow the transfer of air and the numerical model had been developed for vibro-acoustic analysis and flow analysis. The honeycomb panel was improved through the use of corrugation structure in the honeycomb core and two different hybrid honeycomb designs (UCHP and VCHP) had been proposed. The UCHP and VCHP proposed consisted of U-shaped corrugation structure and V-shaped corrugation structure in the honeycomb core respectively. The unweighted STL of the UCHP and VCHP was found to be 19 dB and 25 dB higher than the honeycomb panel with minimal effect on the flow performance. With the improvement in the acoustic properties without significantly affecting the exchange of air, the proposed UCHP and VCHP design could allow the honeycomb panel to be used for wider applications which involved fluid flows such as in motors and compressors, especially when lightweight noise attenuating structure was required. As this paper only focused on the numerical study of the vibro-acoustic characterisation of the honeycomb panel with flow permeability, future work could be conducted to verify the acoustic as well as the flow properties of the panel through experiment verifications.

References

[1] Sui N, Yan X, Huang T Y, Xu J, Yuan F G and Jing Y 2015 *Appl. Phys. Lett.* **106** 171905
[2] Kim S H and Lee S H 2014 *AIP Adv.* **4** 117123
[3] Ma G, Yang M, Yang Z and Sheng P 2013 *Appl. Phys. Lett.* **103** 011903
[4] Huang H, Qiu X and Kang J 2011 *J. Acoust. Soc. Am.* **130** 176-88
[5] P`amies T, Romeu J, Genesc`a M and Arcos R 2014 *Appl. Acoust.* **84** 116-21
[6] Shen C, Xie Y, Cummer S and Jing Y 2017 *J. Acoust. Soc. Am.* **141** 3574
[7] Cho S J, Kim B S, Min D K, Cho Y S and Park J H 2016 *Compos. Struct.* **155** 1-7
[8] Han B, Wang W, Zhang Z, Zhang Q, Jin F, Lu T and Han B 2016 *Theor. App. Mech. Lett.* **6** 54-9
[9] Tang Y, Ren S, Meng H, Xin F, Huang L, Chen T, Zhang C and Lu T J 2017 *Sci. Rep.* **7** 43340
[10] Kim H S, Kim S R, Lee S H, Seo Y H and Ma P S 2016 *J. Acoust. Soc. Am.* **139** 2324
[11] Andersen K S 2008 *Proc. COMSOL Conf.*
[12] Howard C Q and Cazzolato B S 2015 *Acoustic Analysis Using MATLAB and ANSYS* (Boca Raton: CRC Press)
[13] Osipov A, Meesb P and Vermeif G 1997 *Appl. Acoust.* **52** 273-88
[14] Arunkumar M P, Pitchaimani J, Gangadharan K V and Babu M C L 2016 *Aerosp Sci Technol* **56** 155-67
[15] Smolenski C P and Krokosky E M 1973 *J. Acoust. Soc. Am.* **54** 1449-57
[16] Wang T, Li S, Rajaram S and Nutt S R 2010 *J. Vib. Acoust.* **132** 011004
[17] Brosig D, Mellert V and Remmers H 2010 Proc. *20th Int. Congress on Acoustics*

IOP Conf. Series: Materials Science and Engineering **815** (2020) 012008 doi:10.1088/1757-899X/815/1/012008

Transmission Loss Analysis of Simple Expansion Tube with Micro – perforated Cylindrical Panel

MI Alisah[1], LE Ooi[1]*, Z M Ripin[1], CS Ho[2], AF Yahaya[2]

[1]TheVibrationLab, School of Mechanical Engineering, Engineering Campus, Universiti Sains Malaysia, 14300 Nibong Tebal, Pulau Pinang, Malaysia.

[2]Dyson Manufacturing Sdn Bhd, Plot 208, Jalan Cyber 14, Senai Industrial Estate IV, 81400 Senai, Johor, Malaysia.

*Corresponding author: ooiluean@usm.my

Abstract. This study presents a development of simple expansion tube attach with micro-perforated cylindrical panel (MPCP) to improve the acoustic performance. A simulation based boundary element method (BEM) was carried out using PLM Simcenter 3D. The model was constructed in three dimensional in CAD NX Nastran 12 application and then meshing into small elements. The acoustic properties and boundary condition were defined for the simulation purpose. The model was then fabricated by 3D printer material and verified with the transmission loss measurement utilized the two-load method. In comparison, the transmission loss of simple expansion tube with and without MPCP show a good agreement of BEM analysis and experimental result. The addition of the MPCP inside the expansion tube improved the transmission loss by 2-10 dB in wider frequency band compared to the simple expansion tube. Finally, the air cavity depth of the expansion tube is varied to study it effect. Its showed that, the larger air cavity depth caused the transmission loss peak shift to a lower frequency range.

1. Introduction

Simple expansion tubes are widely used in vehicle and industrial for a noise control application. The noise control can be accomplished by applying a porous or fibrous materials inside the muffler [1,2]. However, for the case of high air flux application and clean absorbent, it is not possible to use fibrous or porous materials [3]. A micro-perforated panel (MPP) is a suitable candidate in this kind of application where it can be used inside the simple expansion tube to boost the transmission loss at the desired frequency. Subsequently, MPP is considered to be non-combustible and recyclable, and is suitable for applications in a high temperature environment [4]. Typically, MPP it is made of a metal plate with the perforations diameter in sub-millimeters, and where the air cavity between the plate and the backing wall forms the absorber system [3].

The MPP has been extensively studied by previous researchers. Some have proposed proposed double layer MPP [5,6] and triple layer MPP [7] in order to improve the acoustic performance. However, this flat panel-like shape restricts their practical use in actual rooms or buildings. To overcome these limitations, micro-perforated cylindrical panel (MPCP) have been proposed [8]. The study shows that the noise attenuation of the muffler improved significantly with a broadens frequency after integrating the MPCP.

Transmission loss (TL) is an important parameter for the evaluating the performance of the muffler. TL is studied the discrepancy in the sound power between the incident wave entering and the transmitted wave exiting the muffler when the muffler termination is anechoic [9]. In this study, TL is compute both by simulation through BEM analysis and measurement of two-load method.

The MPCP was integrated inside the simple expansion chamber tube and the result on the sound transmission loss was simulated and validated. The effect of air cavity depth of the MPCP was also studied in order to determine its TL performance.

2. Theoretical Background

TL measurement via two-load approach is accomplished by performing the measurements with two different conditions at the end of the tube, which are rigid and anechoic termination as shown in figure 1. As shown in Figure 1, the sound signal is generated by the sound source at the end of tube, and four microphones are positioned along the tube with the simple expansion tube placed between 2nd and 3rd microphones. The resulting sound field in the tube consists of one component travelling toward the specimen and one reflected component. The sound pressure at the microphone 1, 2, 3, and 4 can be expressed as [3]

$$P_1 = Ae^{[j(\omega t-kl_1)]} + Be^{[j(\omega t+kl_1)]} \quad (1)$$

$$P_2 = Ae^{[j(\omega t-kl_2)]} + Be^{[j(\omega t+kl_2)]} \quad (2)$$

$$P_3 = Ce^{[j(\omega t-kl_3)]} + De^{[j(\omega t+kl_3)]} \quad (3)$$

$$P_4 = Ce^{[j(\omega t-kl_4)]} + De^{[j(\omega t+kl_4)]} \quad (4)$$

where A and B are amplitudes of the incident and reflected sound wave respectively at the inlet tube, C and D are amplitudes of the incident and reflected sound wave respectively at the outlet tube, l_1 is distance from the specimen to microphone 1, l_2 is distance from the specimen to microphone 2, l_3 is distance from the specimen to microphone 3, l_4 is distance from the specimen to microphone 4, k is wave number, and t is the time.

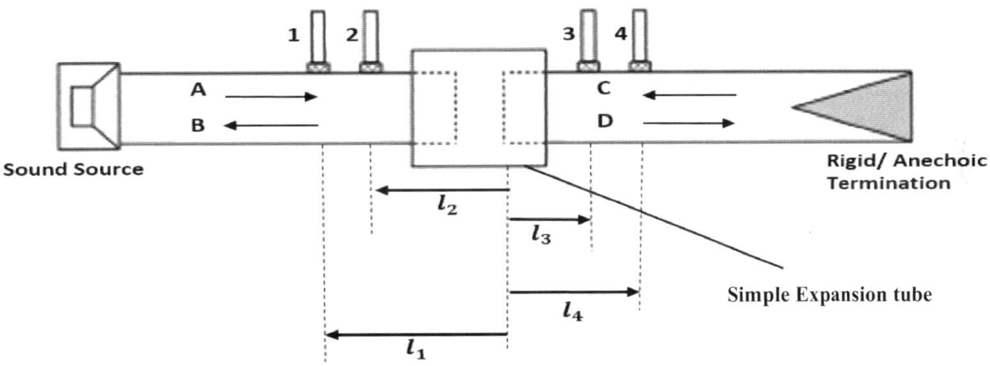

Figure 1. Schematic of two-load transmission loss measurement

The amplitudes of sound wave can be expressed in terms of the sound pressures (P_1, P_2, P_3, P_4) for the four microphones and shown as below

$$A = \frac{j\left[P_1 e^{(jkl_2)} - P_2 e^{(jkl_1)}\right]}{2\sin k(l_1 - l_2)} \tag{5}$$

$$B = \frac{j\left[P_2 e^{(-jkl_1)} - P_1 e^{(-jkl_2)}\right]}{2\sin k(l_1 - l_2)} \tag{6}$$

$$C = \frac{j\left[P_3 e^{(jkl_4)} - P_4 e^{(jkl_3)}\right]}{2\sin k(l_3 - l_4)} \tag{7}$$

$$D = \frac{j\left[P_4 e^{(-jkl_3)} - P_3 e^{(-jkl_4)}\right]}{2\sin k(l_3 - l_4)} \tag{8}$$

There are two sets amplitude of the sound waves are obtained as the two-load method of transmission loss measurement is performed with two different conditions as mention earlier. By inserting and solving the equation (5) to (8), the transmission loss coefficient, τ, for the expansion tube [10] can be calculated by equation below

$$\tau = \frac{A_1 D_2 - A_2 D_1}{C_1 D_2 - C_2 D_1} \tag{9}$$

where A_1, C_1, and D_1 are the amplitude sound wave with rigid termination, and A_2, C_2, and D_2 are the amplitude sound wave with anechoic termination. Finally, the sound transmission loss of the specimen measured in the two-load method can be obtained in unit dB as equation (10) once the TL coefficient is determined.

$$TL = -20\log_{10}(\|\tau\|) \tag{10}$$

An ordinary MPP configuration and its equivalent electro-acoustical circuit is shown in Figure 2. As illustrated in Figure 2, there are four main design parameter in modelling the MPP which are the perforation diameter *d*, the plate thickness *t*, the perforation ratio *σ*, and the air cavity depth *D*.

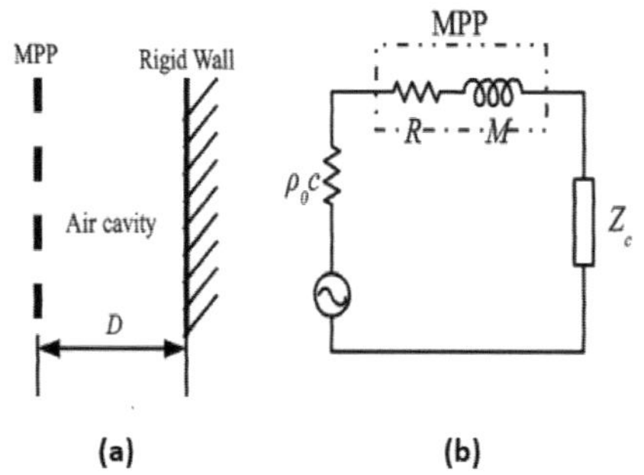

Figure 2. MPP configuration (*a*) electro-acoustical circuit equivalent (*b*) [6]

According to Maa [11], a perforation can be regarded as a short tube, and the specific acoustic impedance of a short tube is expressed as

$$Z_1 = \mathrm{R} + \mathrm{j}\omega\mathrm{M} \tag{11}$$

where

$$R = \frac{32nt}{d^2}\left[\sqrt{1\frac{k^2}{32}} + \frac{\sqrt{2k}}{8}\frac{d}{t}\right] \tag{12}$$

$$\omega M = \omega\rho_0 t\left[1 + \frac{1}{\sqrt{9 + \frac{k^2}{2}}} + 0.85\frac{d}{t}\right] \tag{13}$$

$$k = \frac{d}{2}\sqrt{\frac{\omega\rho_0}{n}} \tag{14}$$

where R and ωM are the specific acoustic resistance and specific acoustic reactance of the perforations respectively. While n is the dynamic viscosity, ρ_0 is the time-average density of air and $\omega = 2\pi f$ is the angular frequency.

3. Methodology

The simple expansion tube is simulated using BEM analysis and then fabricated using 3D printer material based on the layout in figure 3 (a). The main body diameter D_e is 54 mm and 100 mm in length L. The diameters for the inlet d_i and outlet tubes d_o are 34.8 mm. The application of MPCP inside the expansion tube that linked the inlet and outlet tube shows in Figure 3 (b).

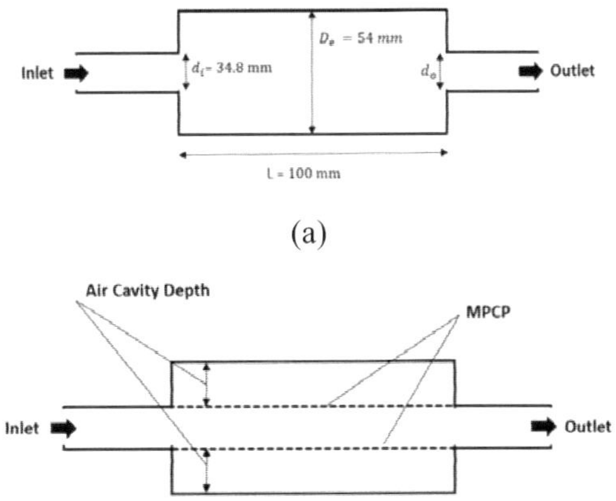

(a)

(b)

Figure 3. Schematic diagram of simple expansion tube a) without MPCP b) with MPCP

The simple expansion tube without and with MPCP is later fabricated by 3D material with dimension as shown in figure 3. A parameter used for MPCP applied inside the simple expansion tube is listed in Design type 1 as shown in table 1. Next, the design of simple expansion chamber with attached MPCP was used for parametric study. The perforation hole diameter, panel thickness and perforation ratio are remaining constant for the parametric study. The effect of the air cavity depth of the expansion tube with MPCP is studied for depth of 6 mm, 9 mm, 12 mm, and 16 mm respectively. Table 1 lists the parameter for four design of the expansion tube.

Table 1. Design parameter for the parametric study of simple expansion tube with attached MPCP

Design type	1	2	3	4
Hole diameter (mm)	0.9	0.9	0.9	0.9
Panel thickness (mm)	1.5	1.5	1.5	1.5
Perforation ratio (%)	1.8	1.8	1.8	1.8
Air Cavity Depth (mm)	6	9	12	16

3.1 Acoustic Simulation and transmission loss measurement

The acoustical simulation model is developed for both model by PLM Simcenter 3D based boundary element method (BEM). The result was then compared with experimental transmission loss measurement for validation. The geometry of the models was constructed in three dimensions in the CAD application environment using NX Nastran and then meshing into small element as illustrated in Figure 4.

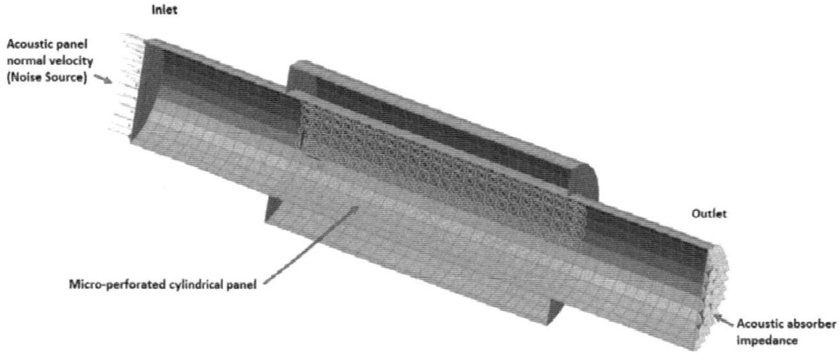

Figure 4. The meshing model of the MPCP in PLM Simcenter 3D.

In this simulation, the sound speed is 340 ms^{-1} and the mass density is 1.225 kgm^{-3}. Two boundary conditions are applied at the inlet and outlet of the expansion tube. The noise source was modelled as the acoustic panel normal velocity with 1 ms^{-1} at the inlet of the expansion chamber [12,13]. The anechoic termination condition at the outlet tube of model can be achieved by inserting the acoustic absorber impedance ($\rho_0 c$) value of 416.5 kgm^{-2} s^{-1} [14]. To compute TL, the sound pressure level of the inlet and outlet tubes of the model was determined based on the point set mesh primitive. The MPCP was defined as an absorbent panel with the impedance values determined based on equation (11) - (14), and fed into the simulation object container as a transfer admittance section of the PLM Simcenter 3D. The transmission loss measurement is based on determination of the transfer matrix using the two-load method as illustrated in Figure 1. Next, transmission loss measurement was set up as illustrated in Figure 1. The measurement set up consisted of four GRAS 46AE microphones, a speaker, a LMS SCADAS data acquisition unit, and a computer with the analysis software which is LMS Test Lab Sound Transmission Loss using impedance tube. The main tube is in the shape of cylinder, with diameter of 34.8 mm and 3 mm wall thickness.

4. Result and Discussion

Figure 5 shows the TL of the simple expansion tube for the frequency from 200 Hz to 5000 Hz. There are two transmission loss curves plotted which are BEM analysis and two-load measurement. The BEM simulation result is represented by black dotted line and solid red line is the measurement result. For frequency range of 400 Hz to 4000 Hz, it is realized that the two transmission loss curves have similar pattern where a dome-shaped with TL peak and valley observed at interval frequency. From this graph, the TL of BEM analysis maintained the same peak level of 3 dB. The transmission loss pattern obtained in this study is comparable and follow a similar trend as a results produced in the literature [14,15]. For frequency of 4000 Hz and above the peaks are slightly higher and shifted to the right for the measurement result. This is because of the inaccuracy of parameters, such as air density as reported in the study of Andersen (2008) [16]. In general, both results from the BEM analysis and two load measurement are comparatively similar and well correlated and BEM approach is considered valid for the simulation of TL of the expansion tube.

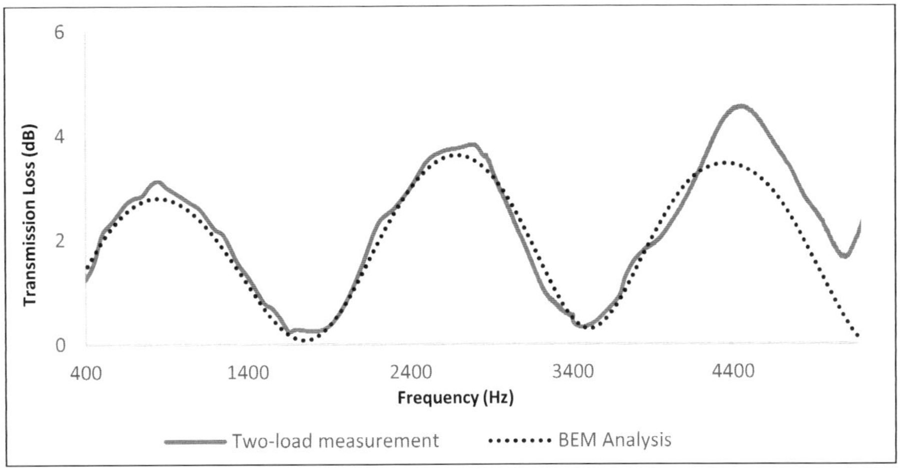

Figure 5. Transmission loss of a simple expansion tube.

Figure 6 shows the TL of BEM analysis and measurement results for both models. From the graph, the BEM analysis is represented by black dotted line curve for simple design, solid red dash line curve for expansion tube with MPCP, solid red line is the measurement result of simple design and solid blue line is measurement result of simple expansion tube with MPCP. It observed that the addition of the MPCP inside the expansion tube leads to the transmission loss improved by 2-10 dB compared to the simple expansion tube. The second hump and third existed with the peak transmission loss value of 12 dB and 14 dB respectively. Above 4000 Hz, it was found that the TL curve became wider and the humps were smoother. In overall, it is found that the transmission loss curve after applying the MPCP produces a higher peak over a wide frequency band compared to the simple expansion tube. The phenomenon is because the micro-perforated inside the expansion tube is playing an important role to increase the noise absorption for the overall investigation frequency range [13].

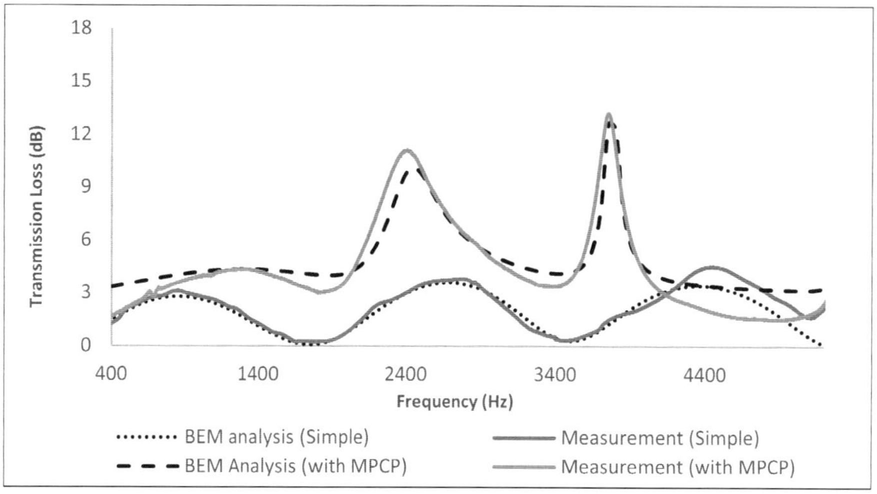

Figure 6. Transmission loss after applying MPCP

Figure 7 shows the parametric study for four different air cavity depth which are 6 mm, 9 mm, 12 mm and 16 mm. The result shows that as the air cavity depth of the MPCP is increased, it will produce higher transmission loss for the frequency range 400-1800 Hz. It is also observed that the peak attenuation of the transmission loss is shifted to low frequency as the air cavity depth value is increased. As reported in the literature by Li [17], the transmission loss pattern obtained in this BEM analysis was comparable the characteristic of the micro-perforated sound absorber, where the larger air cavity depth causes the transmission loss peak to shift to a lower frequency range.

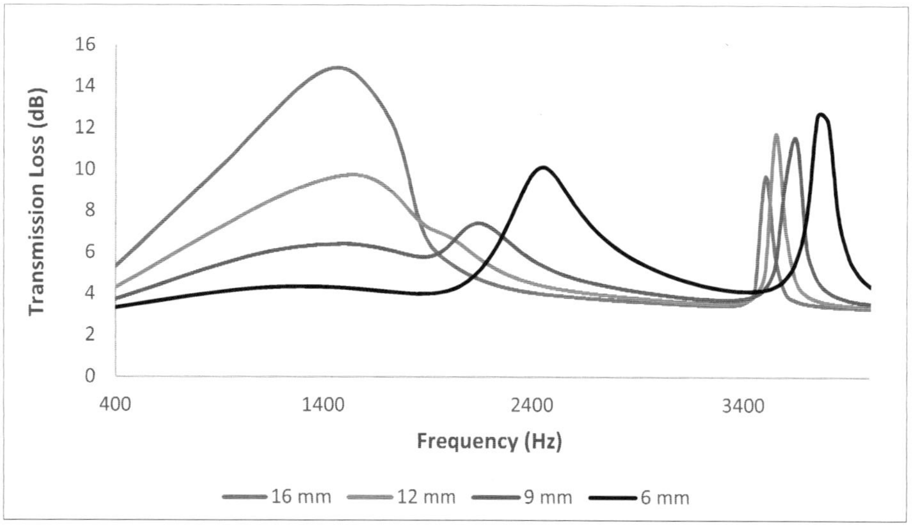

Figure 7. Transmission loss of different air cavity depth of the expansion tube attached with MPCP.

5. Conclusion

The study of the effect of MPCP on the simple expansion tube has done by acoustic simulation using PLM Simcenter 3D and two load method measurement. The result of both methods are comparatively similar and well correlated. The BEM analysis was able to simulate the expansion tube performance accurately for the higher frequency range. From the study, it is found that the addition of the MPCP inside the expansion tube leads to the transmission loss improved by 2-10 dB with a wider frequency band due to effect of MPCP inside the expansion tube. Subsequently, the air cavity depth of the expansion tube also affected the TL curve.

6. References

[1] Delany, M. E., & Bazley, E. N. (1970). Acoustical properties of fibrous absorbent materials. *Applied acoustics*, *3*(2), 105-116.
[2] Na, Y., Lancaster, J., Casali, J., & Cho, G. (2007). Sound absorption coefficients of micro-fiber fabrics by reverberation room method. *Textile Research Journal*, *77*(5), 330-335.
[3] Tan, W. H., & Mohd Ripin, Z. (2013). Analysis of exhaust muffler with micro-perforated panel. *Journal of Vibroengineering*, *15*(2), 558-573.
[4] Wu, M. Q. (1997). Micro-perforated panels for duct silencing. *Noise Control Engineering Journal*, *45*(2), 69-77.
[5] Sakagami, K., Morimoto, M., & Koike, W. (2006). A numerical study of double-leaf microperforated panel absorbers. *Applied acoustics*, *67*(7), 609-619.
[6] Sakagami, K., Nakamori, T., Morimoto, M., & Yairi, M. (2009). Double-leaf microperforated panel space absorbers: A revised theory and detailed analysis. *Applied Acoustics*, *70*(5), 703-709.
[7] Sakagami, K., Yairi, M., & Morimoto, M. (2010). Multiple-leaf sound absorbers with microperforated panels: an overview. *Acoustics Australia*, *38*(2), 76-81.
[8] Sakagami, K., Oshitani, T., Yairi, M., Toyoda, E., & Morimoto, M. (2012). An experimental study on a cylindrical microperforated panel space sound absorber. *Noise Control Engineering Journal*, *60*(1), 22-28.
[9] Tao, Z., & Seybert, A. F. (2003). A review of current techniques for measuring muffler transmission loss. *SAE transactions*, 2096-2100.
[10] Yousefzadeh, B., Mahjoob, M., Mohammadi, N., & Shahsavari, A. (2008). An experimental study of sound transmission loss (STL) measurement techniques using an impedance tube. *Journal of the Acoustical Society of America*, *123*(5), 3119.
[11] Maa, D. Y. (1975). Theory and design of microperforated panel sound-absorbing constructions. *Sci. Sin.*, *18*, 55-71.
[12] Vasile, O. (2010). Transmission Loss Assessment for a Muffler by Boundary Element Method Approach. *Analele Universitatii'Eftimie Murgu'*, *17*(1).
[13] Tan, W. H., & Mohd Ripin, Z. (2013). Analysis of exhaust muffler with micro-perforated panel. *Journal of Vibroengineering*, *15*(2), 558-573.
[14] Tan, W. H., Khor, T. S., & Zunaidi, N. H. (2016). Development of acoustical simulation model for muffler. *International Journal of GEOMATE*, *11*(2), 2385-2390.
[15] Potente, D. (2005, November). General design principles for an automotive muffler. In *Proceedings of ACOUSTICS* (pp. 153-158).
[16] Andersen, K. S. (2008). Analyzing muffler performance using the transfer matrix method. In *Comsol Conference*.
[17] Li, G., & Mechefske, C. K. (2010). A comprehensive experimental study of micro-perforated panel acoustic absorbers in MRI scanners. *Magnetic Resonance Materials in Physics, Biology and Medicine*, *23*(3), 177-185.

Acknowledgements

This study was completed with financial support from the Collaborative Research in Engineering, Science and Technology (CREST) research grant (304.PMEKANIK.6050379). Support from the Universiti Sains Malaysia (USM) and Dyson Development Sdn Bhd is also greatly acknowledged. The authors would also like to express their sincere gratitude to Mr. Baharom Awang and Mr. Wan Muhammad Amri for their assistance in the experimental work in the Vibration Lab.

IOP Conf. Series: Materials Science and Engineering **815** (2020) 012009 doi:10.1088/1757-899X/815/1/012009

The Evaluation of Deviation in Sound Absorption Coefficient for Micro – perforated Panel

N A Jafar[1], L E Ooi[1]*, A Z A Mazlan[1], K H C Seng[2], J M Tan[2]

[1]TheVibrationLab, School of Mechanical Engineering, Engineering Campus, Universiti Sains Malaysia.

[2]Dyson Manufacturing Sdn. Bhd. Plo 208, Jalan Cyber 14, Senai Industrial Estate IV, 81400 Senai Johor, Malaysia

*Corresponding Author: ooiluean@usm.my

Abstract. In this paper, the sound absorption coefficient for micro-perforated panel (MPP) is conducted to determine the acoustic properties of MPP. Two different kind of MPP with the same parameter is used. One is Nylon material and another one is Brass material. Both are fabricated by using 3D Printing technology and Laser Cut process respectively. The sound absorption coefficient of both 3D Printed Nylon MPP and Laser Cut Brass MPP is measured with two-microphone method by using Impedance Tube. The calculated sound absorption coefficient is done by using MATLAB software based on the Classic Maa model. The measured sound absorption coefficient is compared with the calculated result for both MPP. The comparison results show some deviation, like measured results produce a big dip, absorption peak frequency shifted towards higher frequency and wider absorption bandwidth observed when compared to the calculated result. In conclusion, it is found out hole imperfection effect is one of the reasons to the deviation produce between the compared result.

1. Introduction

Micro-perforated panel is first introduced by Maa in year 1975. According to Maa, designer can design the MPP according to their target frequency by optimizing the parameters, such as perforation diameter (d), plate thickness (t), perforation ratio (σ) as well as the air cavity depth (D) [1]. Due to that reason, lots of research has been done from researchers over years studying on the MPP with different design used to suit their objectives, for an example cylindrical shape micro-perforated silencer [2].

In year 2000, Zhang et al., [3] have done a study of micro-perforated panel with different hole diameter (range from 0.17mm to 0.30mm) that aim to produce higher sound absorption. In the study, the authors had derived the equation for plate with two different hole diameter. However, instead of using the derivation to calculate the sound absorption coefficient, the authors calculate two different sound absorption coefficients of MPP with single diameter size (minimum and maximum respectively). Both calculated results are compared with the measured result of MPP with different hole diameter (the diameter size is between the minimum and maximum range). The comparison shows the measurement results gives much wider absorption band frequency compared to the both calculated results even though the absorption peak is slightly lower. In higher frequency range, the measured result looks more closer to calculated result for MPP (with maximum diameter size), but in overall there are some unexplained deviation can be observed along the frequency.

Maury et al., mentioned that it is important to consider the vibration effect of the MPP plate in calculating the sound absorption coefficient of MPP, especially for thin MPP [4]. In the study, two different sound absorption coefficients are calculated. One is taken from Maa that assumed the MPP is rigid and another one is calculated by using the author's derivation for elastic MPP. The calculated results are compared to the experimental result. The comparison between the experimental and the calculated elastic MPP result in overall showed a better agreement compared to the calculated result by Maa model. The experimental result has extra peaks in lower and higher frequency when compared to the calculated results. The authors suggested, the extra peaks are due to the vibroacoustic effect. Thus, the vibration effect needs to be considered when determining the sound absorption coefficient of MPP.

Based on the literature review done, some of the deviations are well explained and some are not properly studied. One of the examples is found in Maury et al. study. The vibration is reported, the vibration effect is neglected in Classic Maa theory [5] due to assumption of rigid MPP [6]. In this study, based on the comparison result between the experimental and calculated sound absorption coefficient result for 3D Printed Nylon MPP and Lased Cut Brass MPP. An investigation will be done to determine the factor that cause the deviation between the comparison result besides vibration effect.

2. Theoretical Method

Micro-perforated panel is a plate that consist of small holes, in sub-millimeter distributed evenly on the plate. MPP works when it is placed in front of a rigid surface with air cavity in between, as shown in figure 1. MPP also can be represent in electrical equivalent circuit, as per shown in figure 1 (b). Sound wave, P_i that impinging on the micro-perforated panel is represented by $2P_i$, the source of sound pressure. While ρc act as the internal air, the mass-resistance element (represent by R and M) of resonance system of the panel is in series with the the cavity reactance of the air cavity depth, Z(D) [7]. The holes act as acoustic mass and the air cavity acts as acoustic spring, which creates a mass-spring resonance relation. Peak of sound absorption occurs when the frequency of sound same with the resonance frequency, causing the stiffness of the air cavity to eliminates the acoustic mass of the holes [8]. The sound absorption by micro-perforate panel always been related to Helmholtz resonator, as the absorption occurs due to the viscous and thermal boundary layer in resonator's neck wall and cavity respectively [9].

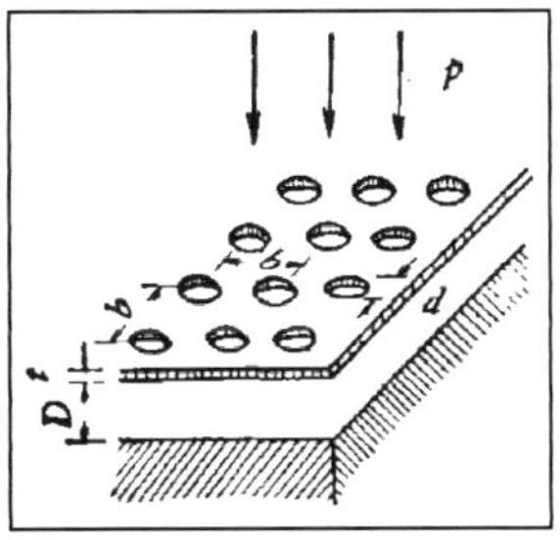

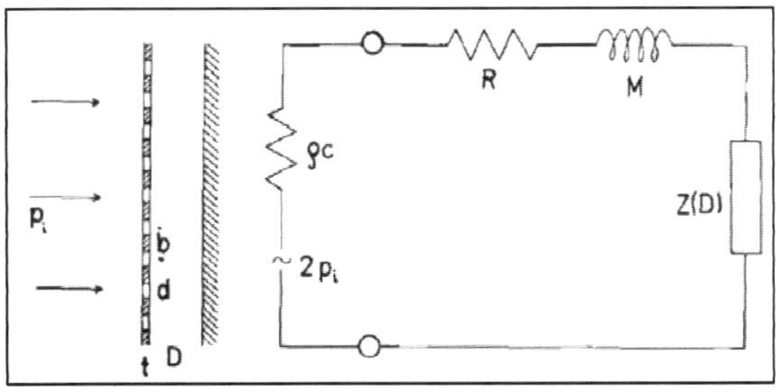

Figure 1. (a) Construction of MPP [10] and (b) MPP electrical equivalent circuit [7]

Firstly, Maa [11] has introduce the specific acoustic impedance equation,

$$Z_1 = j\omega\rho t\left[1 - \frac{2}{k\sqrt{-j}}\frac{J_1(k\sqrt{-j})}{J_0(k\sqrt{-j})}\right]^{-1} \tag{1}$$

The specific acoustic impedance in equation (1) can be represented by the approximate formula below,

$$Z_1 = \frac{32\eta t}{d^2}\left(1 + \frac{k^2}{32}\right)^{1/2} + j\omega\rho t\left(1 + \left(3^2 + \frac{k^2}{2}\right)^{-1/2}\right) \tag{2}$$

The specific acoustic impedance is divided by characteristic impedance of air to obtain the relative acoustic impedance,

$$z = \frac{Z_1}{\sigma\rho c} = r + j\omega m \tag{3}$$

The relative acoustic impedance is divided into real and imaginary part known as relative acoustic resistance, r and mass reactance, m respectively.

$$r = \frac{32\eta t}{\sigma\rho c d^2}k_r \ , \ k_r = \left[1 + \frac{k^2}{32}\right]^{1/2} + \frac{\sqrt{2}}{32}k\frac{d}{t} \tag{4a, 4b}$$

$$\omega m = \frac{\omega t}{\sigma c}k_m \ , \qquad k_m = 1 + \left[1 + \frac{k^2}{2}\right]^{-1/2} + 0.85\frac{d}{t} \tag{5a. 5b}$$

Where, k is the perforate constant, $k = d\sqrt{\frac{\omega\rho}{4\eta}}$. By using equation (4) and (5), we can calculate the normal incidence sound absorption coefficient,

$$\alpha = \frac{4r}{(1+r)^2 + (\omega m - cot\left(\frac{\omega D}{c}\right)^2} \tag{6}$$

Equation (6) will be used later to predict the sound absorption coefficient of 3D Printed Nylon MPP and Lased Cut Brass MPP, then will be compared with the experimental result from Impedance Tube measurement in Result and Discussion section.

3. Methodology

3.1 Samples preparation

Figure 2 shows two MPP samples with same parameter fabricated. One is Nylon material produced by using 3D printing method, another one is Brass material fabricated by using laser cut. Parameter on both MPP are 0.8mm hole diameter with 0.8mm plate thickness, 0.6% perforation ratio and air cavity depth of 4.0mm.

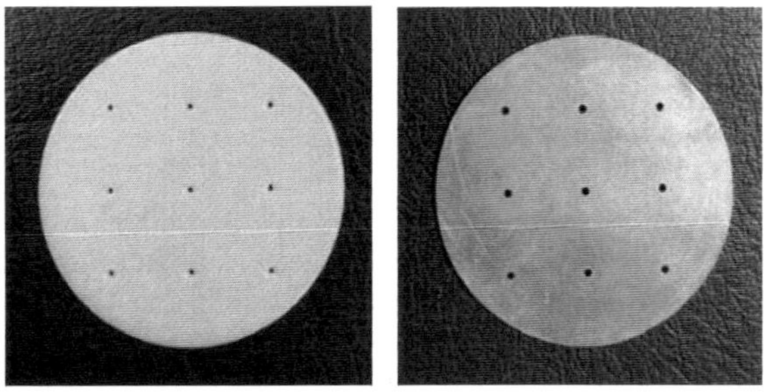

Figure 2. 3D Printed Nylon MPP (left) and Laser Cut Brass MPP (right)

After the samples are fabricated, the sound absorption coefficient is measured by using Impedance Tube according to ISO 10534-2 with LMS Test.Lab software. It is a two-microphone method impedance tube with 34.8mm diameter, able to measure up to 5400Hz frequency. Total of three measurements are done for each MPP samples and the average result is calculated.

3.2 Hole Circularity measurement

The roundness of the hole diameter for each sample is measured by using Optical 3D Surface Metrology Alicona InfiniteFocus. From the measured hole diameter size, the hole circularity is analyzed for both MPP samples. Circularity by using equation (7) is calculated to determine how close the hole to a true circle [12, 13]. While error of circularity by using equation (8) is calculated to analyze the hole quality produce by each MPP samples [14].

$$Circularity, C = \frac{D_{min}}{D_{max}} \tag{7}$$

$$Error\ of\ circularity = R_{max} - R_{min} \tag{8}$$

4. Result and Discussion

4.1 Sound Absorption Coefficient result

Figure 3 shows the comparison in between the measured sound absorption coefficient by Impedance Tube and calculated sound absorption coefficient by Classic Maa model, by using equation (6) for both 3D Printed Nylon MPP and Laser Cut Brass MPP.

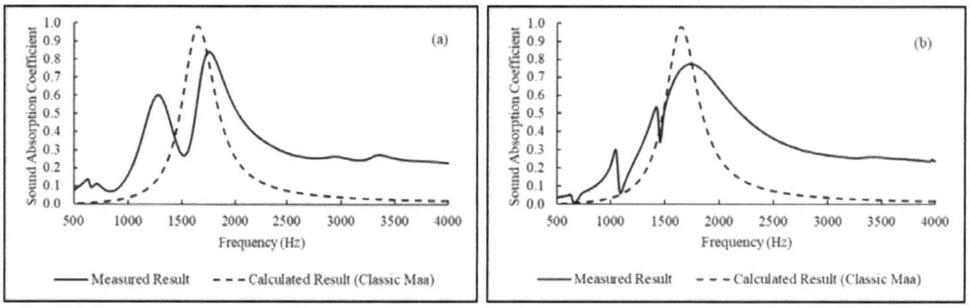

Figure 3. Comparison between measured and calculated (Classic Maa) sound absorption coefficient for (a) 3D Printed Nylon MPP and (b) Laser Cut Brass MPP

Based on observations in figure 3, there are peaks at frequency below 1000Hz. The similar observation is obtained by Chin et al. [8]. The authors explained that it is due to the fundamental modes of the MPP sample and the air resonance occurred inside the impedance tube. For the measured result in figure 3 (a), there is a big dip found at frequency 1518Hz caused by vibrational mode and phase difference between particle and plate velocity [15]. The peak frequency also shifted towards higher frequency for measured result. According to Tan and Ripin [16], the shifted peak is due to the vibration mode shape of the MPP. The similar peak shifts in figure 3 (b) is suspected due to the same reason. Similar observation found in both figure 3 (a) and 3 (b), which is a reduction of sound absorption peak for measured results when compared to the calculated results. The reduction of sound absorption peak is claimed due to the low acoustic resistance of MPP caused by the panel velocity [4].

Based on explanation above, the vibration mode of MPP can cause deviation between measured and calculated results like a big dip and shifting of absorption peak frequency towards higher frequency. However, there is another similar deviation observed in between comparison results in figure 3. Where, both measured results in figure 3 (a) and 3 (b) produce wider sound absorption bandwidth compared to the calculated results by using Classic Maa equation. In the vibration mode of MPP effect, this kind of deviation is not explained. While, in year 2016, Ning et al., [17] reported that a similar result was produced by Wu and Zhao previously. Where the bandwidth of sound absorption coefficient is improved. The authors do a study on the performance of sound absorption for MPP with triangle cross-section perforation shape and burr. The wideband sound absorption coefficient is believed due to the presence of burr that changed the overall hole shape and geometry. However, in Classic Maa theory, the burr effect is never included as straight, constant diameter cylindrical holes are assumed [18].

4.2 Hole Circularity analysis

Continue from the discussion in *4.1*, the deviation resulted from wider absorption bandwidth observed in the comparison results, figure 3 is suspected due to the hole imperfection that are not captured in the Classic Maa theory. Thus, the condition of all hole in both MPP has been observed under InfiniteFocus Alicona Microscope.

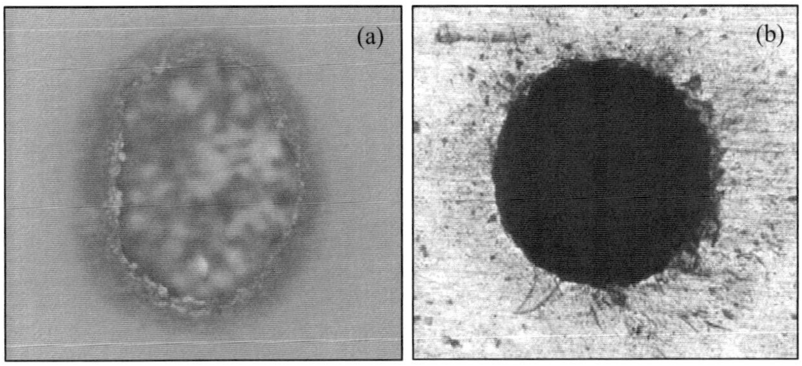

Figure 4. Scanned hole result for (a) 3D Printed Nylon MPP and (b) Laser Cut Brass MPP from InfiniteFocus Alicona Microscope

Figure 4 shows the hole from one of the scanned MPP hole under InfiniteFocus Alicona Microscope. From observation, it is clear that both holes are not are not perfectly circular as per used in theoretical calculation (Classic Maa theory). Mostly the effect of burr that results in non-circular or irregular shape of holes is due to the manufacturing technique [17]. When comparing both results in figure 4 (a) and 4 (b), the hole condition produced by 3D printing technology looks more critical compared to the hole produced by Laser Cut process. Hole of Laser Cut Brass MPP looks more closer to circular shape compared to 3D Printed Nylon MPP, more detail calculations will be done to check on the condition of the hole.

In addition, the measured hole diameter size in 3D Printed Nylon MPP by using InfiniteFocus Alicona Microscope are smaller with range between 0.35826mm to 0.57426mm compared to hole diameter size in Laser Cut Brass MPP with range of 0.90782mm to 1.01430mm. The result of hole diameter size in 3D Printed Nylon MPP can be related to the wide absorption bandwidth observed in figure 3 (a). Same finding is reported by Qian et al., where the reduction in perforation diameter results in good wideband sound absorber [19].

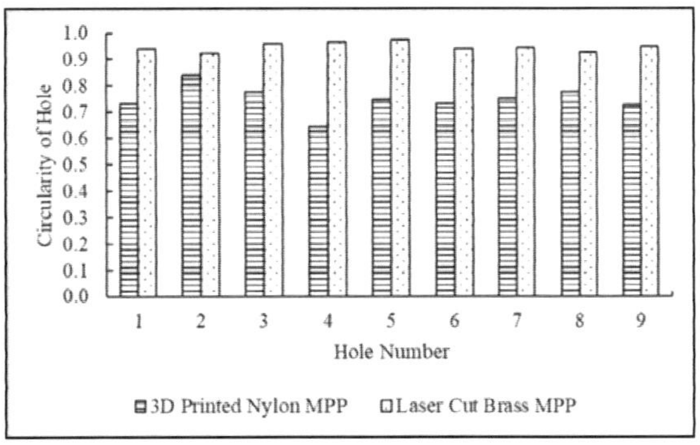

Figure 5. Circularity of Hole result

A detail calculation to explain the condition of holes produce in both MPP is represented by circularity and error of circularity results. By using equation (7), the circularity of each hole in both MPP samples is calculated. There are total of nine holes in each MPP samples. As per judged from the hole circularity results in figure 5, both results are not true circle and not consistent. However, the holes in

Laser Cut Brass MPP are closer to value 1, which means the holes are closer to perfect circular which is equal to 1 [20]. For Laser Cut Brass MPP, each hole circularity is mostly consistence between circularity of 0.9 and 1.0 compared to the holes in 3D Printed Nylon MPP, the circularity results vary between 0.6 to 0.9.

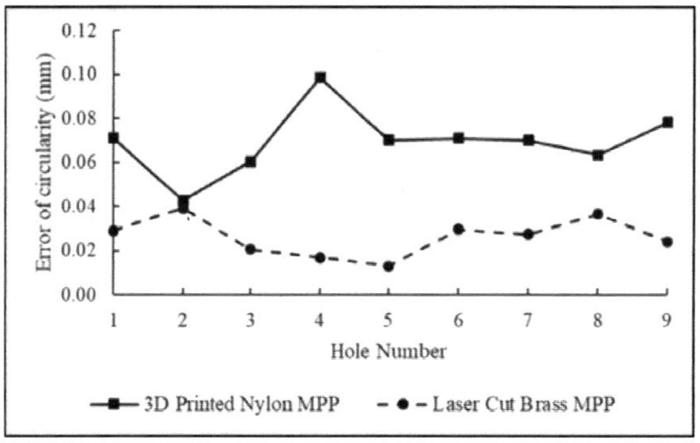

Figure 6. Error of circularity result

From figure 6, it can be observed that the 3D Printed Nylon MPP holes has higher error of circularity compared to the Laser Cut Brass MPP holes. Laser Cut Brass MPP with error of circularity as low as 0.01mm to as high as 0.04mm compared to 3D Printed Nylon MPP with lowest error of circularity at 0.04mm and the highest is recorded a 0.10mm. This result shows that the hole quality produce in Laser Cut Brass MPP is much better compared to the 3D Printed Nylon MPP. In overall, both results in figure 5 and 6 show that condition of hole in 3D Printed Nylon MPP is more critical compared to the hole in Laser Cut Brass MPP as the circularity and error of circularity shown for Laser Cut Brass MPP is closer to 1 and lesser error produce respectively.

All the results obtained from figure 4, 5, and 6 can be related back to the comparison result in figure 3. Imperfect, irregular hole shape and inconsistence hole size produce in both 3D Printed Nylon MPP and Laser Cut Brass MPP results in wider bandwidth frequency observed in figures 3 (a) and 3 (b). The same finding also reported by Ning et al., [17] where it mentions that irregular cross-sectional perforation (like triangle and square shape) gives wider absorption bandwidth compared to regular cross-sectional perforation (like circle shape). The reason is because irregular cross-sectional perforation gives large specific acoustic impedance ratio compared to regular cross-sectional perforation.

5. Conclusion

In this study, the comparison results between measured and calculated sound absorption coefficient for 3D Printed Nylon MPP and Laser Cut Brass MPP is done. Both comparisons show a deviation, which measured results gives wider absorption bandwidth compared to the calculated result by using Classic Maa equation. This is because the scanned hole, circularity and error of circularity result shows that the diameter size of fabricated hole is in range. Where, 3D Printed Nylon MPP records hole diameter size in range of 0.35826mm to 0.57426mm, circularity value varies between 0.6 to 0.9 and error of circularity from 0.04mm to 0.10mm. While, Laser Cut Brass MPP have hole diameter size in range of 0.90782mm to 1.01430mm, circularity value between 0.9 and 1.0 and error of circularity from 0.01mm to 0.04mm. In conclusion, the deviation is found due to the irregular shape and inconsistence size of the perforation hole in the MPP. The reason causing the sound absorption coefficient cannot be calculated correctly as it is not obeyed with the Classic Maa theory, where the hole is assumed to be uniform and perfect circular. Therefore, some correction factors need to be determined for irregular shape and inconsistent perforation hole.

6. References

[1] Maa, D. Y. (1975). Theory and design of microperforated panel sound-absorbing constructions. *Scientia Sinica*, *18*(1), 55-71.
[2] Bravo, T., Maury, C., & Pinhède, C. (2016). Optimisation of micro-perforated cylindrical silencers in linear and nonlinear regimes. *Journal of Sound and Vibration*, *363*, 359-379.
[3] Zhang, B., Lang, S., Ge, P., & Zhuang, W. (2000). The study of sound absorption characteristic of micro-perforated panel with different diameter holes. *Proceedings of Internoise-2000, Nice, France*, 3871-3874.
[4] Maury, C., Bravo, T., & Pinhède, C. (2012, April). Vibroacoustics of thin micro-perforated sound absorbers.
[5] Allam, S., & Åbom, M. (2011). A new type of muffler based on microperforated tubes. *Journal of vibration and acoustics*, *133*(3), 031005.
[6] Li, X., Wu, Q., Kang, L., & Liu, B. (2019). Design of Multiple Parallel-Arranged Perforated Panel Absorbers for Low Frequency Sound Absorption. *Materials*, *12*(13), 2099.
[7] Maa, D. Y. (1987). Microperforated panel wideband absorber. *Noise control engineering journal*, 77-84.
[8] Chin, D. D. V. S., Yahya, M. N. B., Din, N. B. C., & Ong, P. (2018). Acoustic properties of biodegradable composite micro-perforated panel (BC-MPP) made from kenaf fibre and polylactic acid (PLA). *Applied Acoustics*, *138*, 179-187.
[9] Komkin, A. I., Mironov, M. A., & Bykov, A. I. (2017). Sound absorption by a Helmholtz resonator. *Acoustical Physics*, *63*(4), 385-392.
[10] Maa, D. Y. (1985). Wide-band sound absorber based on microperforated panels. *Chinese journal of acoustics*, (3).
[11] Maa, D. Y. (1998). Potential of microperforated panel absorber. *the Journal of the Acoustical Society of America*, *104*(5), 2861-2866.
[12] Bharatish, A., Murthy, H. N., Anand, B., Madhusoodana, C. D., Praveena, G. S., & Krishna, M. (2013). Characterization of hole circularity and heat affected zone in pulsed CO2 laser drilling of alumina ceramics. *Optics & Laser Technology*, *53*, 22-32.
[13] Mustafa, A. Y. (2018). Modelling of the hole quality characteristics by Extreme Learning Machine in fiber laser drilling of Ti-6Al-4V. *Journal of Manufacturing Processes*, *36*, 138-148.
[14] Giasin, K., & Ayvar-Soberanis, S. (2017). An investigation of burrs, chip formation, hole size, circularity and delamination during drilling operation of GLARE using ANOVA. *Composite Structures*, *159*, 745-760.
[15] Tanaka, T., Tatsuhiro, K., & Kazuki, T. (2010). Vibro-acoustic behavior of micro-perforated plate for sound absorption performance. In *20th international congress on acoustics (ICA'10), Sydney*.
[16] Tan, W. H., & Ripin, Z. M. (2016). Optimization of double-layered micro-perforated panels with vibro-acoustic effect. *Journal of the Brazilian Society of Mechanical Sciences and Engineering*, *38*(3), 745-760.
[17] Ning, J. F., Ren, S. W., & Zhao, G. P. (2016). Acoustic properties of micro-perforated panel absorber having arbitrary cross-sectional perforations. *Applied Acoustics*, *111*, 135-142.
[18] Herdtle, T., Stuart Bolton, J., Kim, N. N., Alexander, J. H., & Gerdes, R. W. (2013). Transfer impedance of microperforated materials with tapered holes. *The Journal of the Acoustical Society of America*, *134*(6), 4752-4762.
[19] Qian, Y. J., Kong, D. Y., Liu, S. M., Sun, S. M., & Zhao, Z. (2013). Investigation on micro-perforated panel absorber with ultra-micro perforations. *Applied Acoustics*, *74*(7), 931-935.

[20] Di Ruberto, C., & Dempster, A. (2000). Circularity measures based on mathematical morphology. *Electronics Letters*, *36*(20), 1691-1693.

Acknowledgement

The authors would like to gratefully acknowledge for Collaborative Research in Engineering, Science & Technology (CREST) for the financial funding on this project under accounts (304.PMEKANIK.6050379).

IOP Conf. Series: Materials Science and Engineering **815** (2020) 012010 doi:10.1088/1757-899X/815/1/012010

Oil Palm Wastes as Sustainable Sound Absorbing Particleboard

T K Ong[1], H L Choo[2]*, S M Lee[1] and K Y Kong[1]

[1] Faculty of Engineering and Technology, Tunku Abdul Rahman University College, Kampus Utama, Jalan Genting Kelang, 53300 Kuala Lumpur, Malaysia
[2] School of Engineering, Taylor's University, No 1 Jalan Taylor's, 47500 Subang Jaya, Selangor, Malaysia

*Corresponding author: huileng.choo@taylors.edu.my

Abstract. One method to absorb noise is by using acoustic absorber. Empty fruit bunch fibres (EFB) and mesocarp fibres (MF) are waste materials generated from palm oil mills. This research aimed to fabricate sustainable acoustic absorbing particleboard using EFB and MF to replace conventional materials currently used. Two parameters that were investigated are the mix ratio of the EFB and MF and the addition of binder (urea formaldehyde, UF) into the fibres. Sample thickness was maintained at 20 mm and 30 mm for samples with and without binder. The sound absorption coefficient (SAC) of the samples were tested using an impedance tube according to ISO 10534 – 2 standards. Samples produced using 50% EFB and 50% MF with and without binder shown the best SAC of 0.699 and 0.982 at 1200 Hz respectively. This is equal or better than conventional materials such as mineral wool with SAC values of 0.875 at 1000 Hz. Swelling test however shown that all samples do not meet the criteria by JIS A 5908 for particleboard. Future studies could include the use of green binder and different percentages of EFB, MF and binder.

1. Introduction

Malaysia is one of the top producers of palm oil in the world, accounting to 39% of world production and 44% of world exports [1]. One of the main challenges faced by this industry is the proper disposal of the biomass wastes such as oil palm trunks, empty fruit bunches (EFB), fronds and shells. The implementation of Malaysian Sustainable Palm Oil (MSPO) Certification requires oil palm mills to have a proper waste disposal management programme especially for the waste after the extraction processes. EFB and mesocarp fibres (MF) are the biomass wastes generated after the extraction process. Conventionally, MF are used as feedstock for boiler in palm oil mills to generate electricity locally. However, EFB are underutilised for energy generation and are usually incinerated which causes air pollution [2]. Previous research has shown that natural fibres have the potential to be used as acoustic absorber.

Sound absorber is a material that absorbs sound energy wave generated by the source and through the process, it reduces the noise to the environment. The sound energy is absorbed partly and is turned into other forms of energy. Sound frequency is related to the resonance effect of the material itself. All structures have their own natural frequencies and resonance will occur when the vibrational mode exists similarly. A system can store the energy and transfer it at resonance frequency modes. When the system is working, the process of energy transfer is expected to lose some energy whereby it depends on the damping factor. The study of the resonance phenomena is crucial especially when it comes to sound absorption effect for all kinds of fibres. This is to be able to emphasise on the acoustical/sound properties and able to make changes for noise control applications. Designing of sound absorber depends on the frequency or the damping effect of the material. Besides that, the impedance and sound absorption properties are also affected by the porosity level of the panel itself. For natural fibres sound absorber, it is believed that by mixing different types of fibres with different length and density will change the porosity level of the panel. In order to absorb and impedes higher level of noise, it is believed that the thickness of the sample will affect the sound absorption properties.

In mathematical format, the sound absorption coefficient (SAC), α, can be expressed as equation (1):

$$\alpha = 1 - \frac{I_R}{I_I} \tag{1}$$

where I_R and I_I are the one-sided intensity of the reflected sound and the one-sided intensity of the incident sound respectively. SAC varies from 0 to 1.

The use of oil palm wastes for sound absorption has been studied in the past. Or *et al.* [3] used EFB and investigated the effects of density, thickness and air gap on SAC. They found that 40 mm and 50 mm samples having density of 292 kg/m^3 have the best SAC of 0.9 in average above 1 kHz. The authors also found that the performance of EFB is similar to that of rock wools. Samsudin *et al.* [4,5] studied mixture of EFB and MF with urea formaldehyde (UF) as the binder. EFBMF acoustic panel using 15% of UF as matrix binder achieved 0.58 noise reduction coefficient at a density of 0.4 g/cm^3 [4]. In another research using reverberation room measurement, acoustic panel made from 100% MF coir had the best A rating according to ISO 11654. This means that the panel could absorb more than 90% of incident sound [5]. Istana *et al.* [6] used palm oil fronds and UF as binder and observed to have excellent sound absorption capability. Nasidi *et al.* [7] studied the effects of fibre length and UF content on SAC of EFB. The optimum fibre length was reported to be 2 – 5 mm. In terms of UF content, all samples up to 40% shown good noise reduction coefficient of 0.70 and above with density of around 0.3 g/cm^3.

As research has shown, oil palm wastes have the capability to be used as acoustic absorber. The aim of this paper was to investigate the use of EFB and MF, with and without binder, for use as sound absorbing particleboard according to JIS A 5908.

2. Methodology

2.1. Fibre Preparation

In this research, two types of palm oil residuals, empty fruit bunch (EFB) and mesocarp (MF) fibres, were obtained from Golconda Palm Oil Mill. The obtained fibres were washed thoroughly with water and then immersed in water for 24 hours to remove all the impurities. After that, the fibres were sun-dried for 2 to 3 days. The dried fibres then undergone separation process to remove the kernel shells and other impurities from the fibres.

2.2. Sample Fabrication

The weight of the fibres was measured before placing into the mould for hot compression. For the sound absorber mixed with binder, the weight ratio of binder to fibres was 15:100. The binder used was Urea Formaldehyde (UF) resin, UF 252, obtained from S.A. Wood Chemicals Sdn. Bhd.

A cylindrical mould with a diameter of 100 mm was used to produce the samples. A hot press was used to compress the samples at a temperature of 180 °C and a pressure of 9 MPa for 10 minutes. The amount of EFB, MF and UF used are shown in table 1. For each sample, a total of three specimens were prepared for testing. The density of the samples was calculated by measuring the mass and the volume of each sample and shown in table 1. The density of samples without binder is targeted to be 0.30 g/cm^3 whereas that with binder at 0.50 g/cm^3. The thickness of the samples is remained constant at 30 mm and 20 mm for samples without and with binder respectively.

Table 1. Sample composition and measured average density of three samples.

Sample	Composition			Density (g/cm^3)
	EFB (%)	MF (%)	UF (%)	
100MF	0	100	0	0.28
100EFB	100	0	0	0.27
50EFB50MF	50	50	0	0.30
100MF+UF	0	100	15	0.49
100EFB+UF	100	0	15	0.48
50EFB50MF+UF	50	50	15	0.48

2.3. Sound Absorption Coefficient Measurement
The SAC of the samples was measured using B&K impedance tube kit type 4206 according to ISO10534-2, i.e. the two microphones transfer function method. During the testing, a sample was placed in the opposite end of the sound wave source generator. The measurement frequency was set to 200 – 2000 Hz. Distance between microphones and distance to samples were 50 mm and 210 mm respectively. All necessary calibrations were carried out before starting the measurement.

2.4. Noise Reduction Measurement
Noise reduction measurement was carried out to assess the ability of the sound absorber to reduce noise. It consists of a PVC tube covered with fabric to reduce the effect of noise from the environment. Two covers were made using epoxy to cover the ends of the tube. The white noise source was generated using NCH Tone Generator software with a frequency range 200 – 2000 Hz. Sound level meter, Lutron SL-4011, was used to measure the noise. The background noise generated for this test was 110.8 – 108.7 dBA. Noise reduction is the difference between background noise generated and the measured noise.

2.5. Swelling in Thickness Test
Swelling in thickness test was carried out according to JIS A 5908 for samples with UF binder. This is an important parameter for particleboard. All samples were prepared as before and then cut into 50 x 50 mm cube. The samples were immersed horizontally 30 mm below water surface for 24 hours. The thicknesses of the samples were recorded before and after immersion. The percentage of swelling in thickness after immersion in water is calculated using equation (2).

$$\text{Swelling in thickness after immersion in water (\%)} = \frac{t_2 - t_1}{t_1} \times 100\% \quad (2)$$

where t_1 (mm) and t_2 (mm) are thickness before and after immersion in water respectively.

3. Results and Discussion
3.1. Manufactured Samples
The samples produced are shown in figure 1. As seen in the figure, the samples are of good quality. However, 100MF samples were too soft due to the fine diameter of the fibres and could not be made into shape properly for acoustic testing. Figure 1(a) is the cylindrical sample with a diameter of 100 mm for SAC measurement whereas figure 1(b) are the samples cut into 50 x 50 mm for swelling in thickness test.

Figure 1. Samples of the particleboard for (a) SAC measurement and (b) swelling in thickness test.

3.2. Sound Absorption Coefficient

Figure 2 shows the SAC of different samples for the frequency range 200 – 2000 Hz. Samples made from 100% MF (100MF and 100MF+UF) were not tested as mentioned before. All the samples, with and without binder, show the highest SAC at frequencies between 1200 – 1400 Hz. For samples with binder, the SAC was around 0.694 – 0.699 whereas for that without binder, the SAC was around 0.983. Addition of UF binder lowers the maximum SAC by around 29%. This is because the binder reduces the porosity of the samples, which is essential for sound absorption. Comparing between 100EFB and 50EFB50MF samples, the mixed fibres samples have improved sound absorption capability, increasing the SAC by a maximum of 10% at 800 Hz. As MF fibres have smaller diameter compared to EFB, the compactness of the samples produced enhanced the SAC.

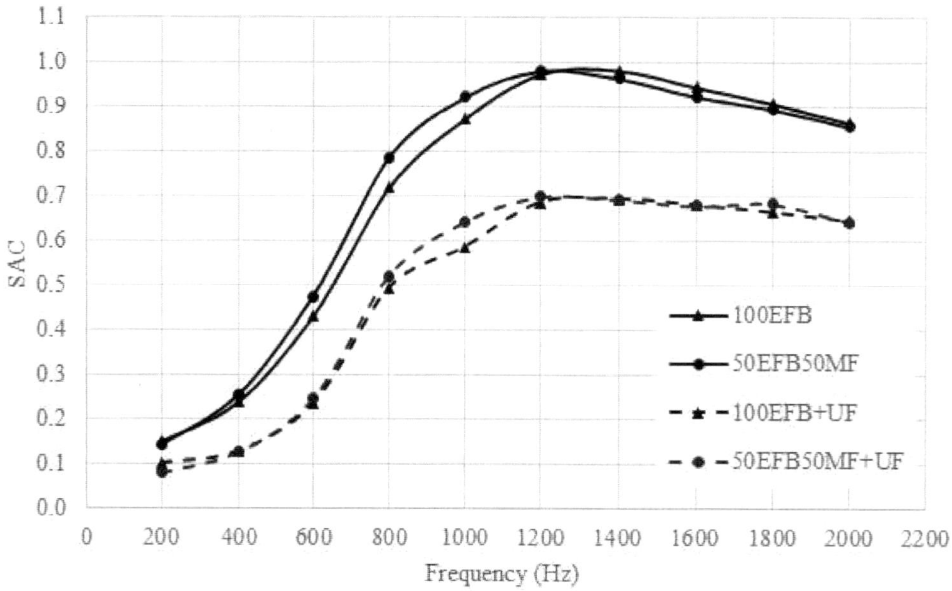

Figure 2. SAC of samples with and without UF binder for the frequency range 200 – 2000 Hz.

The SAC obtained are comparable to or better than synthetic materials such as mineral wool and polystyrene foam as shown in figure 3. Polystyrene foam has the lowest SAC, ranging from 0.041 – 0.514 at 200 – 2000 Hz. Comparing mineral wool and 50EFB50MF, both are comparable up to 800 Hz, after which SAC of 50EFB50MF are higher than mineral wool. Therefore, oil palm wastes have the potential to replace existing material for sound absorption.

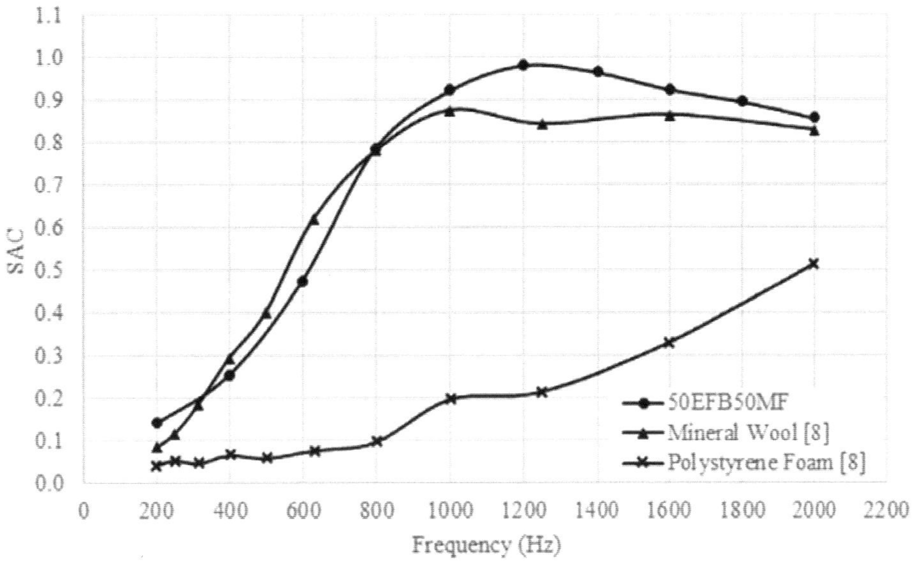

Figure 3. Comparison of SAC for 50EFB50MF, mineral wool and polystyrene foam for the frequency range 200 – 2000 Hz. All samples are 30 mm thick.

3.3. Noise Reduction

Noise reduction results are shown in table 2. All the samples have similar noise reduction range between 5.38-5.85 dBA, around 5% decrease from the generated noise. According to the Malaysian Factories and Machinery (Noise Exposure) Regulations 1989, the permissible exposure time at 85 dBA is 16 hours per day. With every increase of 5 dBA above this, the permissible exposure time is reduced to half. Therefore, a reduction in 5 dBA provided by the samples is seen as significant.

Table 2. Noise reduction in dBA of the samples.

Sample	Noise Reduction (dBA)
100EFB	5.43±0.37
50EFB50MF	5.85±0.56
100EFB+UF	5.45±0.38
50EFB50MF+UF	5.38±0.12

3.4. Percentage Swelling in Thickness

Figure 4 shows the percentage of swelling in thickness after 24 hours immersion in water. Samples with 100% EFB show the highest swelling of around 18% whereas MF only samples at around 12%. The mixed samples gave a value of 15.4%. The swelling in MF samples was the lowest because it has finer fibres, allowing the samples to be more compact. Higher compactness reduces the porosity of the samples. Therefore, water cannot easily penetrate through the bonding between the fibres and the binder.

According to JIS A 5908:2003 for particleboard, the maximum allowable value is 12%. Therefore, the current samples are not suitable for use as particleboard. In the future, addition of additives which can reduce swelling could be investigated.

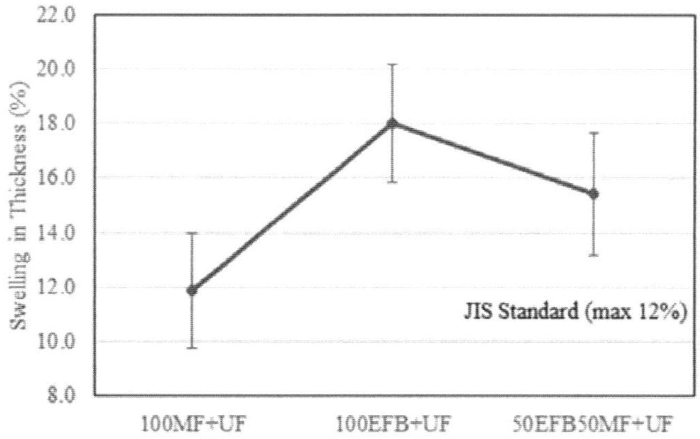

Figure 4. Percentage swelling in thickness after immersion in water for samples with UF binder.

4. Conclusions

This paper aimed to produce sound absorbing particleboard from oil palm wastes. The results obtained have shown that palm oil wastes can be utilised as efficient sound absorbing materials, comparable to or better than the synthetic materials currently used. However, the samples did not pass the swelling in thickness test, which is an important parameter for particleboard. In the future, different ratios of EFB, MF and UF could be investigated. Other green binders could also be looked into to replace UF, resulting in a truly sustainable and green sound absorbing particle board.

5. References

[1] Malaysian Palm Oil Council 2019 Malaysian Palm Oil Industry - MPOC

[2] Abdullah N and Sulaiman F 2013 The Oil Palm Wastes in Malaysia *Biomass Now - Sustainable Growth and Use* ed M D Matovic (IntechOpen)

[3] Or K H, Putra A and Selamat M Z 2017 Oil palm empty fruit bunch fibres as sustainable acoustic absorber *Appl. Acoust.* **119** 9–16

[4] Samsudin E M, Ismail L H, Abd Kadir A and Nasidi I N 2017 Thickness, Density and Porosity Relationship Towards Sound Absorption Performance of Mixed Palm Oil Fibers *24th International Congress on Sound and Vibration* (London)

[5] Samsudin E M, Ismail L H, Abd Kadir A, Nasidi I N and Samsudin N S 2018 Rating of Sound Absorption for EFBMF Acoustic Panels according to ISO 11654:1997 ed S A Aljunid, M A A Mohd Salleh, C B M Rashidi, P J Soh and K N F Ku Azir *MATEC Web Conf.* **150** 03002

[6] Istana B, Ridwan A, Badrun Y and Retnawaty S F 2018 Utilization of Waste Stem of Oilpalm For The Manufacture of Particle Board Composite Sound Absorbency *International Conference CelSciTech-UMRI 2018*

[7] Nasidi I N, Ismail L H, Samsudin E M, Abdul Khodir M F and Kamarozaman M A 2018 The Effect of Different Fibre Length and Different Urea Formaldehyde (UF) content on Sound Absorption Performance of Empty Fruit Bunch (EFB) ed S A Aljunid, M A A Mohd Salleh, C B M Rashidi, P J Soh and K N F Ku Azir *MATEC Web Conf.* **150** 03003

[8] Sikora J and Turkiewicz J 2010 Sound Absorption Coefficients of Granular Materials *Mech. Control* **29** 149–57

IOP Conf. Series: Materials Science and Engineering **815** (2020) 012011 doi:10.1088/1757-899X/815/1/012011

Sound Absorption Characteristics of Sandwich Panel Made from Double Leaf Micro-perforated Panel And Natural Fiber

Meifal Rusli*, Ronal Satria Nanda, Hendery Dahlan, Mulyadi Bur

Department of Mechanical Engineering, Faculty of Engineering, Andalas University, Kampus Limau Manis, Padang 25163 – Indonesia

*Corresponding email: meifal@eng.unand.ac.id

Abstract. One of sound-absorbent developed nowadays, micro-perforated panel (MPP), effectively reduce reflected acoustic energy after approach the panel based on a Helmholtz-type resonance absorber. Some MPPs modifications have been applied to increase its performance, like the application of double-leaf MPP. Furthermore, the implementation of sound absorption materials made of natural fibers recently has invited more attention to some researcher because it is biodegradable, environmentally friendly, and economical. In this paper, a sandwich panel made of double MPP and natural fiber as sound absorptive material is investigated. The MPPs are made of a transparent acrylic board 1.5 mm thickness, 0.8 mm hole's diameter with five mm hole distance in front panel, and five mm in the backing panel. Two different types of natural fibers panel; pineapple leaf fiber and oil palm fiber from empty fruit bunches are used in this sandwich. a low-cost impedance tube with double channel microphone is used to measure the sound absorption coefficients. It is found that the sandwich model changes the sound absorption characteristics of MPP by shifting the maximum absorption coefficient into the lower frequency and raising the absorption coefficient of the double MPP following its natural fiber panel absorption coefficient. The total impedance of sandwich gives the effect of decreasing the frequency peak of the absorption, and higher resistance of the fiber panel will increase the absorption coefficient and broaden the frequency peak.

1. Introduction

Concerning noise, its attenuation is becoming a significant issue nowadays. It is considered as one of the pollutants that not only from the annoyance point of view but also health problems and quality of life. The first way to control the noise is by attenuating it from the source by considering all possible noise generations during the designing process of a product. Unfortunately, so far, this design process has not entirely successfully eliminated the noise from its source. In this situation, the possible solution is to reduce the noise on its transmission path and the receiver side.

Sound-absorbent materials are then generally used to minimize the noise on its transmission path by reducing reflected sound hitting a hard or rigid interior wall or surface. The material absorbs some acoustical energy, and remain less part then is reflected. Various of sound absorptive materials are now available and developed. Recently, there have been two methods that commonly used and developed as the sound absorber panel, i.e. porous material and micro-perforated panel (MPP) [1,2]. Porous absorptive material is mostly categorized based on its microscopic structures; fibrous, cellular, and granular. Rock wool, glass wool, polyester, and polyurethane are fibrous materials that generally found in this application. Nevertheless, Their process of manufacture and application contributes to some unfavorable impacts to human and environmental health [3].

IOP Conf. Series: Materials Science and Engineering **815** (2020) 012011 doi:10.1088/1757-899X/815/1/012011

Therefore, researchers then give more attention to developing the use of natural fibrous material as sound-absorbent, because they are biodegradable, naturally renewable, and environmentally friendly. Several kinds of natural fibers have been investigated and developed as sound absorptive material, like palm oil male flower [4], kenaf, cork, hemp, cotton, wood, cane, cardboard, and sheep wool [5], the combination of rice husk grain and coconut coir [6], extracted pineapple-leaf fibres and empty fruit bunches of oil palm fiber [7]. Some researchers also developed natural fiber-reinforced composites to enhance the lack of the natural fiber's mechanical and moisture properties [8, 9].

Furthermore, micro-perforated panels (MPPs) are investigated being able to absorb the sound energy effectively by composing of a big number of small Helmholtz resonators. While the sound waves enter an MPP, the sound energy is converted into the heat by the friction among the perforation surfaces and the moving air molecules. The idea of MPP was introduced first as an alternative for porous sound absorbers that remains the health and environmental problems [10]. However, MPPs are still working in specific frequency, and low absorption outside the resonance frequency range [11].

Various methods have been developed by many researchers to improve the frequency band, and sound absorption coefficient of MPP, such as inserting a mechanical impedance plate behind the panel [12], partitioning different depth cavities [13], combining with a structure consist of a membrane cell and mass blocks composite [14], and multi-layering elastic MPP plates [15]. Moreover, other researchers also proposed improvement by making the MPP from composite materials, such as wooden panels [16], kenaf and polylactic acid [17], and hybrid MPP combination with perforated date palm panel and porous layer [18].

Besides, pineapple leaf and empty fruit bunches of oil palm are two of the natural fiber source that abundantly available in many tropical countries. Pineapple leaf fiber, for example, has been used in some traditional applications like textile and home accessories. Because it's some good mechanical properties, the fiber from oil palm also has a good chance to be developed commercially [14].

In this paper, the sound absorption characteristic of a sandwich panel made of a natural fiber material inserted between a pair of MPP panel is investigated experimentally. MPPs are made from a thin transparent acrylic panel with different sizes and density of the micro-hole. Two different types of natural fibers panel; pineapple leaf fiber and empty fruit bunches fibers of oil palm are used in this sandwich.

2. Measurement Method

2.1 Material

The fibers from pineapple leaf and empty fruit bunches of oil palm, firstly, are cleaned by spraying with mineral water and then drying naturally at room temperature. The sample panels are made in a 54 mm diameter cylindrical shape following the impedance tube diameter. The fiber panel is composed using a dies using a short PVC pipe. The thin layer of fiber is added into the dies, and a paper glue is dabbed by brush between the layers. The specimen is then compressed to obtain 20 mm of thickness. The glue then let dry naturally at room temperature. The density of the samples is shown in Table 1. Figure 1 shows the natural fiber panels and MPP. Pineapple leaf panel has more density than the palm empty fruit bunches one.

Table 1. The natural fiber panels density

The sample material	Density (gr/cm^3)
Palm empty fruit bunches	0.2328
Pineapple leaf	0.3279

Then, two MPPs are made of a transparent acrylic panel with 1.5 mm thickness, 0.8 mm hole's diameter and three mm distance between the holes in backing panel (MPP A), and 0.8 mm hole's

IOP Conf. Series: Materials Science and Engineering **815** (2020) 012011 doi:10.1088/1757-899X/815/1/012011

diameter and five mm distance between the holes in front panel (MPP B). It means that the perforation ratio is about 0.056% for backing panel and 0.02% for the front panel. The sandwich panel is then made from two MPPs, and a natural fiber panel is inserted between the MPPs panel.

Figure 1. Fiber panel samples; pineapple leaf fiber panel (left), and empty fruit bunches fiber (middle) and acrylic MPP (right).

2.2. Experimental setup

A low-cost impedance tube with double microphones with a 54 mm inner diameter made of aluminum pipe is used to conduct the measurements within a frequency range of 200 Hz to 3000 Hz like illustrated in Fig. 2. The testing set up is designed based on ASTM E 1050. The load speaker generates chirp sound within a frequency range of 100 Hz to 10 kHz as a sound source. The sound pressure is measured by two BSWA microphones MPA 215 type with each sensitivity factor 43.7 Pa/mV and 51.2 Pa/mV. These two microphones are then connected to the NI 9234 data acquisition. Firstly, the impedance tube calibration is conducted using Rockwool 45 with a density of 45 kg/m^3 and 40 mm thickness. The coefficient of sound absorption is then obtained by the transfer function of two microphone channels. The sample panel composition set-up is depicted in Fig. 3.

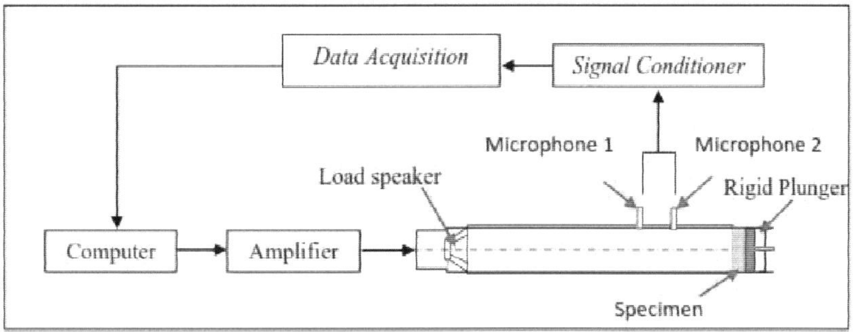

Figure 2. Double microphones channel impedance tube set up.

There are three compositions of panels reported in this paper, as illustrated in Fig. 3. The first is fiber panel only (Fig. 3a), double leaf MPPs where MPP A is set to make five mm air gap in front of the rigid plunger and the MPP B is allocated 20 mm in front of MPP A and directly faces the sound source (Fig. 3b), the sandwich panel by inserting fiber panel between the MPPs. The change of the sound absorption coefficients between the various set-up they will be observed.

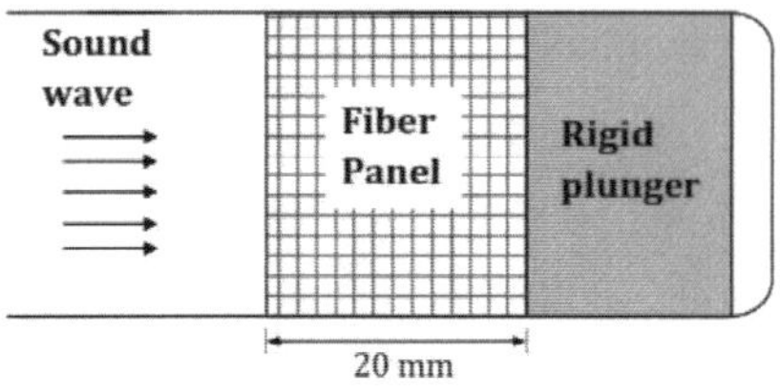

a. The fiber panel testing set up

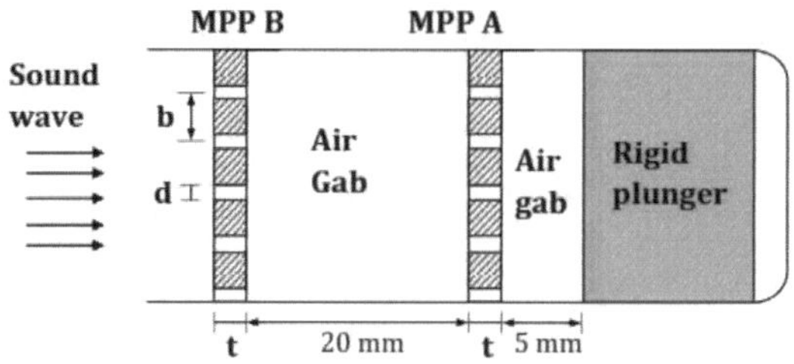

b. The double leaf MPPs testing set up

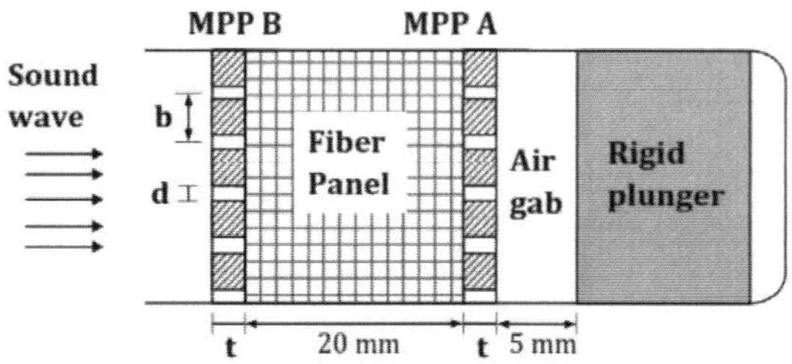

c.The sandwich panel testing set up

Figure 3. The testing panel setup in impedance tube

3. Result and Discussion

The sound absorption coefficient of both fiber panels is depicted in Fig. 4. The sound absorption characteristic of pineapple leaf fiber, in general, is seen higher than empty fruit bunches of oil palm, especially at up to 3000 Hz observed frequency range. The maximum absorption coefficient approach 100% at frequency of 2200 Hz. On the other hand, empty fruit bunches panel only has up to 93% sound absorption but shows a tendency to increase or keep constant at higher frequency. Moreover, both panels shows very low absorption coefficient at less than 500 Hz, and some noisy signal appears in this impedance tube measurement.

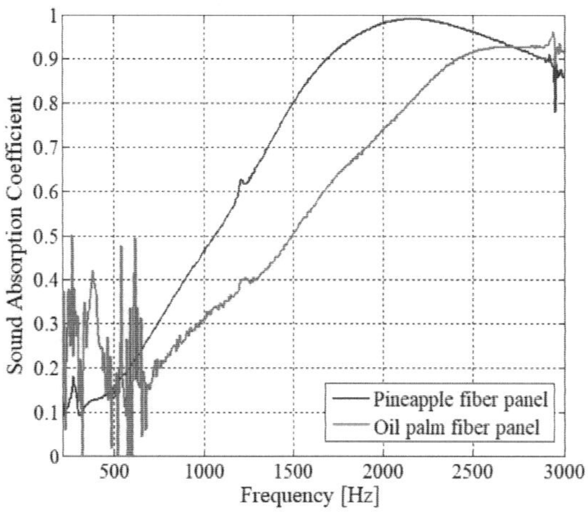

Figure 4. The sound absorption characteristic of the fiber panel

It is found that material density contributes more effect on the absorption properties. Higher fiber density and thickness of the panel increase its absorption coefficient. More fibers available make more energy of sound pressure trapped within the panels and raise the absorption coefficient. The fiber size of both natural sources observed under the microscope are depicted in Fig. 5. It is shown that the empty fruit bunches oil palm fiber has larger diameters, more empty volume, and low density. On the other hand, pineapple leaf fiber is shown has uniform fiber size and higher density. The fiber diameter and density will affect the ability of the fiber panels to absorb the acoustic wave energy that hits its surface. Moreover, More dense fiber in panel increases the resistance of airflow by the viscosity friction through the air molecule's vibration. Therefore, when a sound wave enters the fibers, the wave amplitude is reduced by friction like the waves trying to move through pass the tortuous passages. Therefore, the acoustical energy then is converted into heat [1].

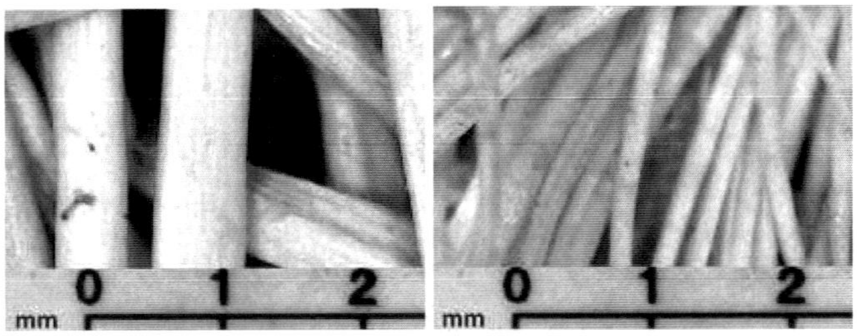

Figure 5. The fiber size of empty fruit bunches oil palm (left) and pineapple leaf (right)

Then, the sound absorption coefficients of MPP and palm empty fruit bunches fiber panel are depicted in Fig. 6. Double leaf MPPs in this model have a deficient ability to absorb the sound energy, only up to on the characteristics of MPPs themselves, air cavity between the leaves, and air gab backing the MPP [19]. However, in this paper, the MPPs are setting in constant. On the other hand, the palm fiber panel may absorb more than 93% at 2500 Hz and more. The sandwich panel reduces the frequency peak into 1000 Hz, with sharp peaks and a maximum 93% sound absorption.

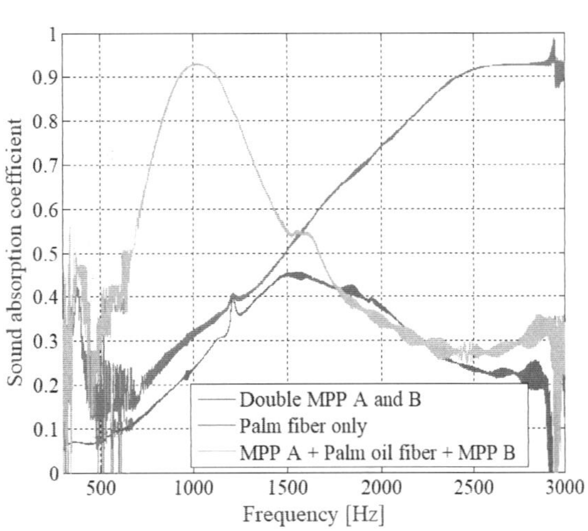

Figure 6. The sound absorption characteristic of MPP and empty fruit bunches fiber sandwich panel.

Likewise, the sound absorption coefficients of MPP and its sandwich panel with pineapple leaf fiber are shown in Fig. 7. The pineapple leaf fiber panel has an almost 100% sound absorption coefficient at 2200 Hz. The sandwich panel reduces the frequency peak into 800 Hz, with very sharp peaks and a maximum 100% sound absorption.

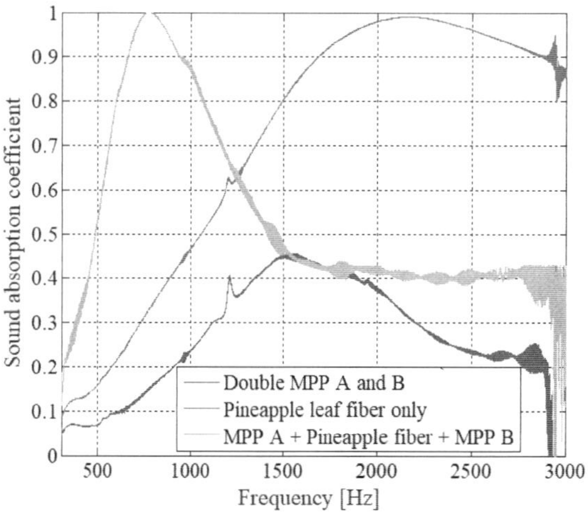

Figure 7. MPP and pineapple leaf fiber sandwich panel sound absorption characteristics.

Firstly, a multiple-leaf MPP idea was proposed by Maa [10] using a double-leaf MPP and a rigid-back wall behind the air-cavity. This model produces two resonators to get a wider range of frequency absorption. Nevertheless, when MPP has a back wall, sound absorption occurs only by the Helmholtz resonance, and the absorption is restricted in the region of its resonance frequency [19]. In some conditions, the sound absorption is not only limited in a sharp frequency range but also produces a lower absorption coefficient.

Previously, in general, it was known that the fiber panels, including natural fibers, have good sound absorption coefficients at a higher frequency. Furthermore, a wider gap of air cavity MPP has a lower frequency peak of the sound absorption. In Fig. 6 and 7 can be observed that the sandwich panel of both using two types of natural fiber increases the absorption capacity at a lower frequency. It means that the frequency peak of the absorption coefficient is shifted into a lower frequency. Therefore, when the acoustic resistance of the double-leaf MPP is not good at low frequency, inserting the fiber panel will be significant to increase the absorption performance.

When the fiber panel that kind of a porous layer is inserted in the air cavity between the MPPs, the absorption may be deteriorated by the large acoustic resistance due to the porous absorbent. The effect is not only to increase the acoustic resistance, but also to the impedance. The total impedance of front MPP, fiber panel, backing MPP, and air cavity give the effect of decreasing the frequency peak of the absorption. When the resistance of the fiber panel is higher at a wider frequency band, it may be expected that a fiber panel can broaden the frequency range of absorption of the sandwich panel by the additional damping by the porous absorber.

4. Conclusion

The measurement and analysis of the sound absorption coefficient of sandwich panels made from the double-leaf micro-perforated panel and pineapple leaf and empty fruit bunches of oil palm fiber have been done. From the result and discussion, it is found that the sandwich model of double-leaf MPP inserted by natural fiber changes the sound absorption characteristics of MPP by shifting the maximum absorption coefficient into the lower frequency and increasing the absorption coefficient of doble-leaf MPP into the maximum absorption in accordance to the maximum absorption coeeficient of its natural fiber panel. The total impedance of sandwich gives the effect of decreasing the frequency peak of the absorption, and higher resistance of the fiber panel will increase the absorption coefficient and broaden the frequency peak.

5. References

[1] Seddeq H S 2009 Factors Influencing Acoustics Performance of Sound Absorptive Materials *Australian Journal of Basic and Applied Science* 3(4), pp 4610-4617.

[2] Arenas J P and Crocker M J 2010 Recent Trends in Porous Sound-Absorbing Materials *Sound & Vibration* July 2010, pp 12-17

[3] Asdrubali F, Schiavoni S and Horoshenkov K V 2012 A Review of Sustainable Materials for Acoustic Applications *Building Acoustics* 19(4) p 283–312.

[4] Masrol S R, Rosdin M K R and Ibrahim M N 2013 Sound absorption characteristics of palm oil male flower spike fiber-reinforced composite *International Conference on Mechanical Engineering Research* July 2013, pp 1-9.

[5] Berardi U and Iannace G 2015 Acoustic characterization of natural fibers for sound absorption, *Building and Environment* 94-2 pp 840-852.

[6] Mamtaz H, Fouladi M H, Nuawi M Z, Namasivayam S N, Ghassem M and Al-Atabi M 2017 Acoustic absorption of fibro-granular composite with cylindrical grain *Applied Acoustics* 126, pp 58–67

[7] Rusli M, Irsyad M, Dahlan H, Gusriwandi, Bur M 2019 Sound absorption characteristics of the natural fibrous material from coconut coir, oil palm fruit bunches, and pineapple leaf *IOP Conf. Series: Materials Science and Engineering* 602, 012067

[8] Jayamani E and Hamdan S 2013 Sound Absorption Coefficients Natural Fiber Reinforced Composites *Advanced Materials Research* 701, pp 53-58.

[9] Suresha K.V, Shivanand H K, Amith A, Vidyasagar N 2018 Evaluation of mechanical properties of hybrid fiber (hemp, jute, kevlar) reinforced composites *AIP Conference Proceedings* 1943(1):020109,

[10] Maa D Y 2007 Practical single MPP absorber *International Journal of Acoustics and Vibration* 12, pp 3-6.

[11] Sakagami K, Kobatake S, Kano K, Morimoto M and Yairi M 2011 Sound absorption characteristics of a single microperforated panel absorber backed by a porous absorbent layer *Acoustics Australia* 39, pp 95-100.

[12] Xiaodan Z and Xiangqian F 2015 Enhancing low frequency sound absorption of micro-perforated panel absorbers by using mechanical impedance plates *Applied Acoustics* 88, pp 123–128.

[13] Guo W, Min H 2015 A compound micro-perforated panel sound absorber with partitioned cavities of different depths *Energy Procedia* 78, pp 1617 – 1622.

[14] Gai X L, Xing T, Li X H, Zhang B, Wang F and Cai Z N 2018 Sound absorption properties of microperforated panel with membrane cell and mass blocks composite structure *Applied Acoustics* 137, pp 98–107.

[15] Kim H S, Ma P S, Kim S R, Lee S H and Seo Y H 2018 A model for the sound absorption coefficient of multi-layered elastic micro-perforated plates *Journal of Sound and Vibration* 430, pp 75-92

[16] Song B, Peng L, Fu F, Liu M and Zhang H 2016 Experimental and Theoretical Analysis of Sound Absorption Properties of Finely Perforated Wooden Panels, *Materials* 9, 942.

[17] Chin D D V S, Yahyaa M N, Dinb N B C and Onga P 2018 Acoustic properties of biodegradable composite micro-perforated panel (BC-MPP) made from kenaf fibre and polylactic acid (PLA) *Applied Acoustics* 138, pp 179–187.

[18] Elwaleed A K, Nikabdullah N, Nor M J M, Tahir M F M and Zulkifli R 2013 Experimental investigation of sound absorption properties of perforated date palm fibers panel, *IOP Conf. Series: Materials Science and Engineering* 46, 012027.

[19] Sakagami K, Morimoto S and Yairi M 2007 Recent development in the application of microperforated panel absorbers *14th Internasional Congress on Sound and Vibration* July 2007 Cairns Australia, pp 1-18.

Acknowledgments

The authors gratefully acknowledge the financial support from the Mechanical Engineering Department, Faculty of Engineering, Andalas University in the year of 2019.

IOP Conf. Series: Materials Science and Engineering **815** (2020) 012012 doi:10.1088/1757-899X/815/1/012012

A Qualitative Study of the Wind Noise Cavity Inside of a Two Way Radio

W M Hairudin[1], N M Kamarudin[1], L E Ooi[2], N A Ismail[1]*

[1]School of Aerospace Engineering, Engineering Campus, Universiti Sains Malaysia, 14300 Nibong Tebal, Pulau Pinang, Malaysia.
[2]School of Mechanical Engineering, Engineering Campus, Universiti Sains Malaysia, 14300 Nibong Tebal, Pulau Pinang, Malaysia.

*Corresponding author: aenorilmi@usm.my

Abstract. The cavity inside microphone of two-way radio can induced noise when involved in outdoor and windy condition. An investigation was performed in Viwika wind tunnel involving a flow visualizations techniques to clarify the flow noise behavior generated by the wind, particularly within the cavity. A smoke wire test was conducted using a cavity model at varying aspect ratio (L/D) between 0.7 and 3 in angle of attack of 0° with Reynolds number of 7831. A two-dimensional Computational Fluid Dynamic (CFD) models based on k-Epsilon turbulence model were conducted to validate the flow visualization result. The results revealed that the deep cavity with an aspect ratio (L/D) of 0.7 produced a single clockwise vortical flow structure and the aspect ratio (L/D) of 3 show the complex flow structure which two recirculation flow vortices were produced inside the cavity. The flow behavior in the CFD simulation was similar to the smoke test. It is shown the flow behaviour inside cavities was control by the aspect ratio (L/D).

1. Introduction

Wind noise is known as one of the issues in mobile telecommunication such as two-way radio. In this application, communications involve the use of a microphone inside of two-way radio. When two-way radio are used in outdoor activities and windy condition, wind noise interferes with the signals, particularly in the low-frequency range. The wind noise existed due to the wind that passes the microphone's cavity structure and vibrating the diaphragm, creating a signal that equal to or greater than the voice receiver. High wind flow affects the quality of the signal receiver due to the presence of turbulent flow over the microphone's cavity and contaminates the desired output signal. Thus, it severely degrades the quality and intelligibility of the speech signal [1].

The phenomenon of wind noise has been studied by numerous applications in the past such as in landing gear [2], car sunroof [3], vehicles [4-5], high speed train (pantograph) [6 -7], building [8-9], closed side branches of pipelines [10], wind turbines [11], submarine hulls [12], musical instruments [13] and microphones [14-15]. The microphone's cavity is a typical kind of structure that can induced noise. The flow cavity noise rises from the feedback mechanism between the shear layer instabilities and acoustic waves that cause an aeroacoustic resonance. The aeroacoustic resonance also leads to fatigue and vibration in structures. The aeroacoustic cavity resonant can be divided into three groups;(1) Rossiter resonance which includes the high-speed Mach flow over the shallow cavities (L/D >1) [2], (2) the cavity resonance by Helmholtz resonator or deep cavity or side branch [16-17] and (3) the acoustic resonance inside the corrugated or coaxial pipe [18-20]. The unsteady oscillations of the air often caused due to disturbance, such as sharp edges of a rigid structure causing turbulence or vortices. The turbulence and vortices produced when the large pressure fluctuates by the impinging a shear layer on the leading edge (A) of the cavity. The shear layer flow is separated from the leading edge of the cavity and breaks down into downstream, and the emerging vortex shedding impinging on the trailing edge (B) results in pressure waves. The pressure waves lead to a self-sustained oscillation, inducing acoustic noise as shown in Figure 1. The self-sustained oscillations flow over a cavity can radiate tonal noise as a cavity tone.

One of the parameters that influence the flow behaviour and oscillation modes within the cavity is cavity geometry which includes an aspect ratio (L/D). Yilmaz *et al.* [21] conducted a 2D simulation in various L/D in ANSYS Fluent. The result shows that by increasing the L/D, the flow field within cavities become complex and have multiple modes. In L/D of 1, there is no pressure oscillation due to the less energy in shear layer. Yoshida *et al.* [22] investigated the flow field and oscillation mode of varied cavity length to depth ratio (L/D). The result revealed that flow field is influenced by the relationship of cavity shear layer oscillating mode and recirculating vortices within the cavity. The oscillating mode can be switch into three groups which includes non-oscillating mode, shear layer and wake mode.

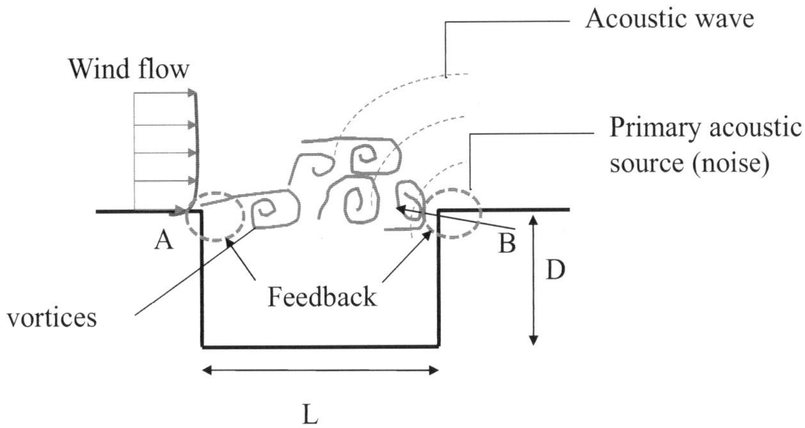

Figure 1: Mechanism of cavity flow

The wind induced noise inside two-way radio has been studied by several researchers. Fisol *et al.* [23] has investigated the effect of the flow speed and the angle of attack with noise level in open wind tunnel. They concluded that the noise level is increased when flow speed and angle of attack increases. Another work by [24] performed a sound measurement in term of threshold hearing level intelligibility, angle of orientation and wind speed of the signal receiver. The results have shown that the angle leads to worst wind-induced noise effect is at 90° to the leading edge, which is showing the highest sound pressure level. Yow [ ] performed CFD simulation and validate the result with actual prototype model. They result shown that. Based on this review, not much attention into flow behaviour and acoustic analysis were taken into the cavity of two-way radio. Thus, it is important to determine the flow behaviour that causes the noise inside the cavity and clarify the method to suppress the cavity resonant.

In this paper, the flow behavior is focused on the open cavity inside the two-way radio with a different aspect ratio of (L/D) were qualitatively studied by smoke visualization technique and compared with CFD simulation. The research results provides the reference for reducing the noisy speech signal for the two-way radio during windy environments.

IOP Conf. Series: Materials Science and Engineering **815** (2020) 012012 doi:10.1088/1757-899X/815/1/012012

2. Flow visualisation technique

2.1 Experimental set-up

Experiments of flow visualization were performed in small scale closed loop wind tunnel. The wind tunnel has a dimensions of 0.2m (width) x 0.4m (height) and 0.8m (length). The maximum achievable speed is 11 m/s.

2.2 Cavity model

Two cavity models with different L/D were designed in Solid work and and printed using in house 3D Expresso printer at the Space System Laboratory at Universiti Sains Malaysia. The cavities are open cavity with different aspect ratio of L/D = 0.7 and 3 as shown in Figure 2. The depth was kept fixed. All cavities model were printed using black PLA material to reduce the light reflection during experiment.

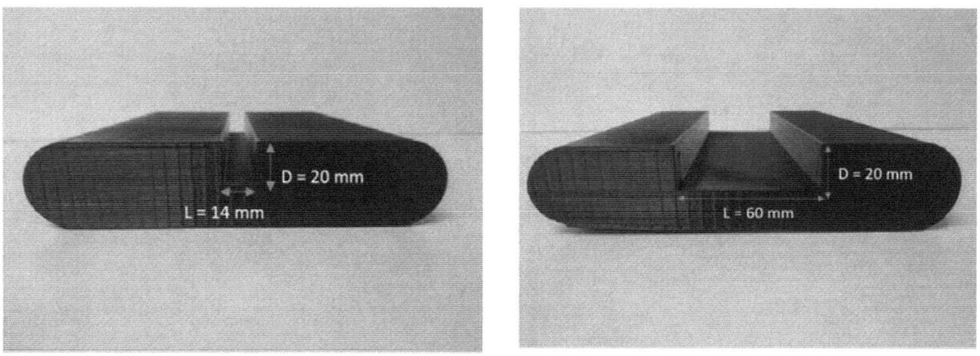

(a) (b)

Figure 2: 3D printed test model with two cavities at length to depth ratio (a) L/D = 0.7 (b) L/D = 3

2.3 Experimental procedure

A picture of wind tunnel set up is shown in Figure 3(a). The cavity model was mounted on the axis in the centre of the measurement section. The axis has an outer diameter of 6.2mm and the alignment bolt is 2mm in diameter. The different model cavity of length to depth ratio (L/D) of 0.7 and 3 were tested with Reynolds number of 7831 corresponding to the speed of 0.8m/s at angle of 0°. A video camera with 30 frames per second (fps) was located outside the test section door to capture the snapshots of the flow. All the images were recorded and analyzed. However, in this paper only clear image of vortex flow is illustrated. The door test section was covered with black cloth to avoid any reflection. The velocities at the inlet and inside the test section was measured by a portable digital hot wire anemometer.

In this experiment, water was used as a working fluid to produce the smoke line. The water was installed in the tank up to a level within the Min - Max indicators as shown in figure 3(b). The waterproof ultrasonic vaporizer was used to generate the fog in the wind tunnel. An ultrasonic vaporizer uses a ceramic diaphragm vibrating at an ultrasonic frequency to create water droplets that exit the fog inlet in the form of cool fog. The fog then gets forced out by mini fan blower that placed upper the working fluid tank. The fog flows to the tunnel and started to flow along smoke inlet by using fan blower. The fan blower was used to reduce the turbulence and straighten the flow only in axial direction. A heat grid (hot wire) generates the streamlines in the test section.

IOP Conf. Series: Materials Science and Engineering **815** (2020) 012012 doi:10.1088/1757-899X/815/1/012012

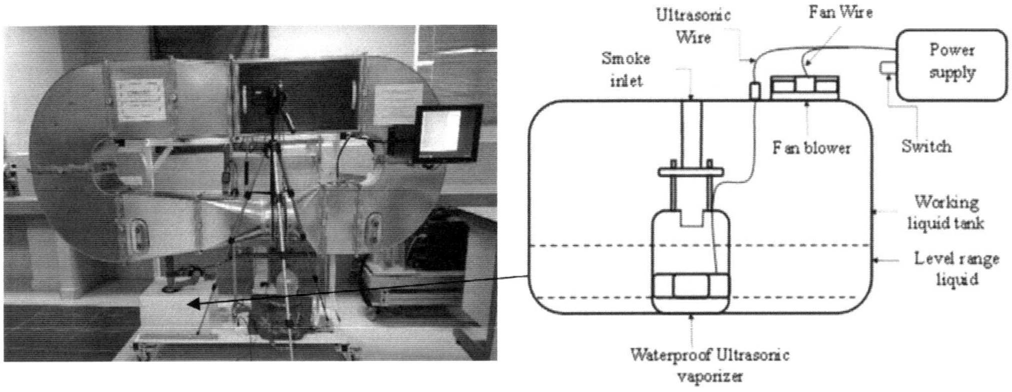

(a) (b)

Figure 3: (a) Viwika close loop wind tunnel (b) Schematic diagram of water tank in Viwika Wind Tunnel

3. Computational Fluid Dynamic (CFD)

The commercial CFD software Ansys – FLUENT 16.0 [23] was used to simulate the flow field inside the small cavity**.** The simulation was based on the 2-dimensional (2D) using Reynolds Averaged Navier-Stokes (RANS) equation approach. A k-epsilon turbulence model was chosen to simulate the flow characteristics due to the shear layer, recirculating flow and small pressure gradients as reported by [25]. A semi-implicit method for pressure-linked equations (SIMPLE) algorithm of the pressure-velocity coupling was selected to solve the governing equations for low speed incompressible flow. The flow conditions were specified as Re = 7831 corresponding to U = 0.8 m/s. Table 1 shows the condition for the simulation parameter of the cavity model. The CFD structured meshes of each cavity shown in figure 4 meanwhile the nodes and element were summarized in Table 2. The computational time step of the transient flow field was 0.01s at number of step size of 350. Simulation parameters were tabulated in Table 1. Figure 5 shows the rake location point for velocity profile in ANSYS Fluent.

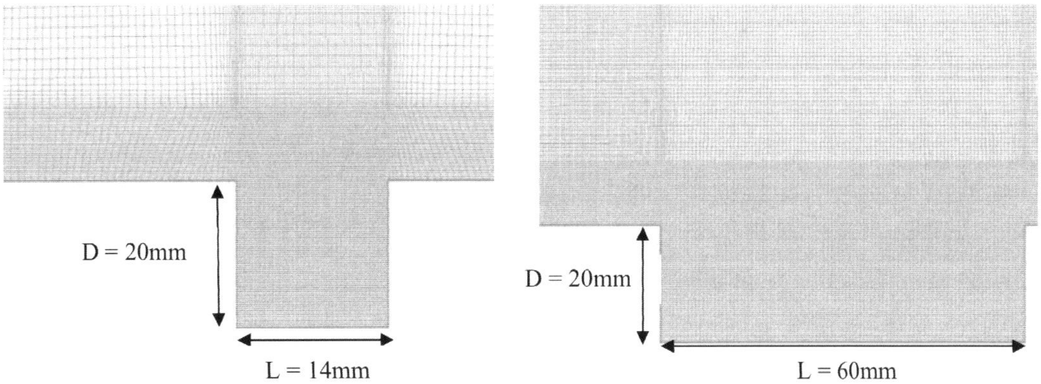

(a) (b)

Figure 4: The structured mesh of different cavity model (a) L/D =0.7 (b) L/D =3

Table 1. Simulation parameters of the cavity model.

Simulation parameter	Value
Air density, p (kg/m^3)	1.225
Kinematic viscosity, υ	1.7 x 10^{-5}
Pressure, P (Pa)	101325
Time step size (s)	0.01

Table 2. Nodes and elements of each CFD meshes.

Cavity model	Nodes	Elements
Cavity model (L/D= 0.7)	41860	41371
Cavity model (L/D = 3)	57885	57324

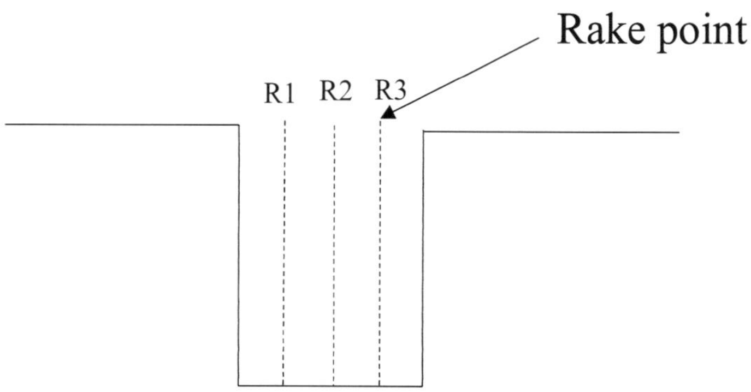

Figure 5: The rake point for velocity value in CFD simulation

4. Mesh Independence Study

The effect of the mesh size can be seen in Figure 4. In this figure, the blab la show close agreement with each other. The coarse mesh influence the ability to resolve the

5. Result and Discussion

In this section the flow behaviour inside cavity were presented. The effect of different of aspect ratio (L/D) were discussed details in term of the velocity contour.

5.1 Effect of aspect ratio (L/D)

The analysis on the experiment and computational results indicate that some vortex generates from flow separation in the shear layer inside the cavity as shown in Figure 6. The flow structures of the effect of aspect ratio (L/D) are presented in Figure 6(a) and (b). From the figures, it can be seen as the aspect ratio (L/D) increase, the flow structure inside cavity also affected. In aspect ratio L/D of 0.7, the flow separates at the leading edge of the cavity and impinges at the trailing edge. A single vertical flow structure was present within the cavity. For L/D of 3, the flow seems reattached at the floor of the cavity downstream before the flow separating again at upstream leading edge. It can be observed that for L/D =3, two counter rotating vortices present within the cavity due to more energy to the forming shear layer, compared with L/D=0.7 that has less interaction with forming shear layer near the leading edge [26]. It also can be seen the rake point (R2) in Figure 7 was the highest velocity compared the others rakes in both aspect ratio (L/D). This result revealed that R2 is the area of oscillating vortex flow mode.

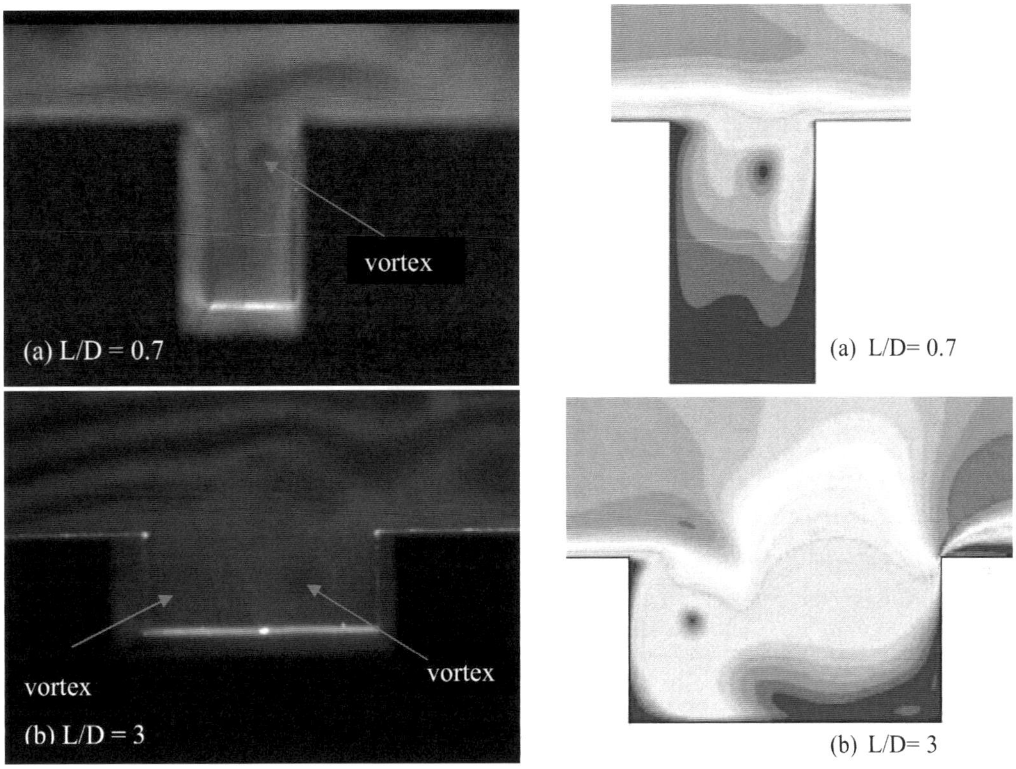

Figure 6. Comparison result of the flow structure at Reynolds number of 7831 between smoke test experiment and CFD simulation for different aspect ratio (L/D): (a) L/D = 0.7 and (b) L/D = 3

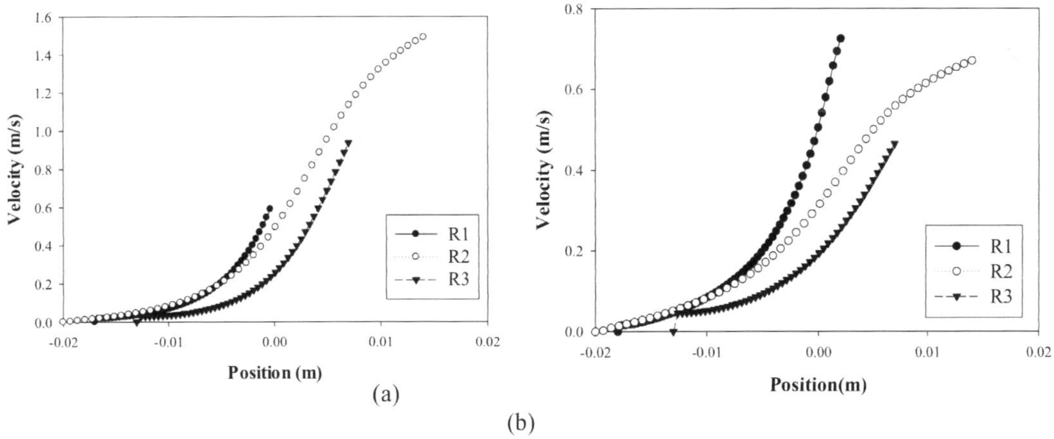

Figure 7: The velocity profile inside cavity of two-way radio at varied length to depth ratio (L/D): (a) L/D: 0.7 (b) L/D: 3

6. Conclusion

The flow structure inside different aspect ratio (L/D) of the cavity were investigated by experimentally and numerically. The present study concludes that, as the cavity length changes from 0.7 to 3, the flow structure become complex and the secondary vortex flow inside the cavity becomes larger. However, the snapshot of secondary clockwise vortex inside a larger aspect ratio was not clearly captured due to the poor resolution by the camera. In length to depth ratio (L/D) of 0.7, only one continuous recirculation vortex flow occurred within the cavity. It was also found that the velocity of the vortex shed in the middle of the cavity (R2) at length to depth ratio (L/D=0.7) was the highest velocity compared with the other point rake location and length to depth of 3. The comparison of flow structure in CFD and smoke test were in good agreement. The smoke wire flow visualization performed inside the Viwika wind tunnel was able to visualise the vortex flow inside cavity. However, the smoke lines were big due to the heat grid size inside wind tunnel. The effect of aspect ratio in flow structure was important parameter that causes an oscillating flow mode. Further work is needed to measure the oscillation flow mode inside the cavity using microphones.

7. References

[1] Xu Y, Zheng ZC, Wilson DK 2011 A computational study of the effect of windscreen shape and flow resistivity on turbulent wind noise reduction *J Acoust Soc* **129(4)** 1740-7.
[2] Neittaanmäki P, Rossi T, Korotov S, Oñate E, Périaux J, Knörzer D 2004 Landing gears aerodynamic interaction noise," *Eur. Congr. Comput. Methods Appl. Sci. Eng* 115–35.
[3] Wang Y, Deng Y, Yang Z, Jiang T, Su C, Yang X. 2013 Numerical study of the flow-induced sunroof buffeting noise of a simplified cavity model based on the slightly compressible model. *J Auto Eng.* **8** 1187-99.
[4] Chen KH, Johnson J, Dietschi U, Khalighi B 2010 Wind noise measurements for automotive mirrors. *SAE Int. J. Passeng. Cars - Mech. Syst* **2.** 419–33.
[5] Alam F, 2000 The Effects of Car A-pillar and Windshield Geometry on Local Flow and Noise
[6] Thompson DJ, Latorre Iglesias E, Liu X, Zhu J, Hu Z. 2015 Recent developments in the prediction and control of aerodynamic noise from high-speed trains. *Int. J of Rail Trans.* **3** 119-50.
[7] Sun, X. and Xiao, H 2018. Numerical modeling and investigation on aerodynamic noise characteristics of pantographs in high-speed trains. *Complexity.*
[8] Čeheľová D, Urbán D, Bielek B, Rychtáriková M, Roozen NB 2017 Literature review on wind induced sound on buildings. *Akustika.* **1** 42-7.
[9] Wei Xu and Feng Xu, "Numerical Study on Wind-Induced Noise of High-Rise Building Curtain Wall with Outside Shading Devices," Shock and Vibration. **12**
[10] Mohamed, S Graf, H.R. and Ziada, S 2012 Aeroacoustic Source of a Shallow Cavity in a Pipeline. In *ASME 2011 Pressure Vessels and Piping Conference, Ame Soc of Mech Eng Dig Col* 269-276
[11] Gallo P, Fredianelli L, Palazzuoli D, Licitra G, Fidecaro F. 2016 A procedure for the assessment of wind turbine noise. *Applied Acoustics* Dec 15; **114** 213-17.
[12] Zeng, W.D., Wang, Y.S. and Yang, Q.F., 2010. Numerical calculation of flow noise of submarine with full appendages. *Acta Armamentarii*, ***9.***
[13] Fabre B, Gilbert J, Hirschberg A, Pelorson X. 2012 Aeroacoustics of musical instruments. *Annual review of fluid mechanics* **44:**1-25
[14] Wang, L Zander, A.C. and Lenchine, V.V 2012 Measurement of the self-noise of microphone wind shields. *Measurement, 3* 7.
[15] Staple, B.W., Schladt, G.A. and Holmes, M.E., Motorola Solutions Inc, 1994. Wind noise and vibration noise reducing microphone. *U.S. Patent* **5**, 288,955.
[16] Amandolese X, Regardin C 2004 An experimental study of the acoustic oscillations by flows over cavities. J. of Vib and Acou. 126(2) 190-5.
[17] Ghanadi F, Arjomandi M, Cazzolato B, Zander A. 2014 Interaction of a flow-excited Helmholtz resonator with a grazing turbulent boundary layer. *Exp. Therm. and Flui Sci.* **58**:80-92.

[18] Meissner M 2002 Acoustic modes induced by flow in a pipe with two closed side-branches. *App. Acous*. Oct 1; **63(10)**:1071-83.
[19] Tonon D, Hirschberg A, Golliard J, Ziada 2011. Aeroacoustics of pipe systems with closed branches. *Int. J. of Aero* **10(2-3)** 201-75.
[20] Slaton WV, Nishikawa A 2015 Aeroacoustic response of coaxial wall-mounted Helmholtz resonators in a low-speed wind tunnel. *The J. of the Acous. Soc* 37**(1)** 253-60.
[21] Yilmaz I, Ayli E, Aradag S 2013 Investigation of the effects of length to depth ratio on open supersonic cavities using CFD and proper orthogonal decomposition. *The Scie World J*
[22] Yoshida T, Watanabe T. 2016 Numerical simulation of flow over an open cavity with self-sustained oscillation mode switching. *Open J of Fluid Dynamics* **6(04)** 361.
[23] Fisol UM, Ripin ZM, Ismail NA, Saad AA, Hashim MK, Chan PY, Yow HT, Tan CH, Hii UH, Saw TS 2013 Wind noise analysis of a two-way radio. *IEEE International Conference on Smart Instrumentation, Measurement and Applications (ICSIMA) 1-5*
[24] Hashim MK, Ismail NA, Saad AA, Chan PY, Fisol UM, Ripin ZM, Yow HT, Tan CH, Hii UH, Saw TS 2013 Sound quality analysis for two-way radio under wind noise. *IEEE International Conference on Smart Instrumentation, Measurement and Applications (ICSIMA)* 1-5.
[25] ANSYS Fluent version 16 Documentation
[26] Fisol UM, 2016 Numerical and Experimental Analysis of Two – Way radio, Master Thesis.

Acknowledgement
The authors acknowledge this research work was fully funded by the Research Universiti Sains Malaysia under RUI Grant (1001/PAERO/8014005). Thanks to technical supports, Mr Mahmud Isa from School of Aerospace Engineering for the help.

IOP Conf. Series: Materials Science and Engineering **815** (2020) 012013 doi:10.1088/1757-899X/815/1/012013

Attenuation of Motorcycle Handle Vibration using Suspended Handlebar

Nabil Mohamad Usamah[1] and Zaidi Mohd Ripin[1]*

[1]TheVibrationLab, School of Mechanical Engineering, Universiti Sains Malaysia,14300 Nibong Tebal, SPS, Pulau Pinang, Malaysia

*Corresponding author: mezaidi@usm.my

Abstract. Motorcycle rider is exposed to the handle vibration. Extended exposure to high level of handle vibration can result is hand-arm vibration syndrome (HAVs). In South East Asia the majority of the motorcycle are powered by single cylinder small displacement (70 – 125 cc) engines. This paper describes the modal analysis of a small single cylinder underbone type motorcycle widely used in the rural and urban for commuting to workplace. A suspended handlebar is developed by mounting the handlebar on rubber mounts. Transmissibility measurement are carried out for static and on the road condition. There are conditions where transmissibility is above unity due to resonance for the existing handlebar and this condition is overcome by the suspended handlebar except at frequency of 25 Hand this was below the operating frequency range of the motorcycle engine of cruising speed (50 – 60 Hz). On the road tests show that the handlebar is effectively attenuate with the installation of the engine mounts. Maximum attenuation of 79 % is achieved at 3300 – 3400 rpm which give HAV of 3.41 ms^{-2}.

1. Introduction

Motorcycles are important type of transportation especially in rural areas and high traffic road. Motorcycle riders are easily exposed with high level of vibration transmitted from the motorcycle handlebar in certain cases can be greater than the limit value set by the European Directive (2002/44/EC). Methods must be found to reduce vibration if the vibration exposure is greater than the action limit value of 2.5 ms^{-2}. The Directive also requires that the vibration exposure to be less than 5.0 m/s2. The International Standard ISO 5349 [1] & [2] suggests that a person who is exposed to vibration at the vibration exposure limit value and vibration action limit value will have a 10% chance of getting finger blanching after 5.8 years and 12 years of vibration exposure, respectively. In 1996 and 2016, the vibration on the motorcycle handle is reported in the range from 2.2 to 4.9 ms^{-2} [3] and from 3.82 to 9.77 ms^{-2} [4] at the lower range of frequency spectrum. The an anti-vibration gloves have been proven to be general ineffective especially in the lower weighted acceleration spectrum [5]. The prevalence of finger blanching is found to be at 4.2% among police traffic motorcyclists in Japan [3], indicating that the vibration transmitted to the hand poses a serious long term health risks. This includes the disorder of the vascular, muscle, neurological and bone and joint. Another aspect of motorcycle handle vibration is the discomfort to the rider. These are usually based on perceptions and many survey-based studies have confirmed the influence of vibration on the comfort of the rider. A study showed that over 50% of respondents (both female and male respondents) reported hand and arm discomfort whereas motorcycle operation will always be accompanied by vibration, which will make the rider and his passenger feel unpleasant[8] & [7]. A study based survey on has shown that musculoskeletal disorders are prevalent among the occupational motorcyclist and most of the vibration came from the handlebar and the foot rest [8], [9].

The study presented here is to reduce the vibration transmission from the motorcycle to the rider hands by installing a suspended handlebar. The approach is to carry out pre-evaluation using the laboratory transmissibility test. To characterize and study the effect of varying stiffness and damping values of the suspended handlebar, structural dynamic modification (SDM) is used to determine the maximum transmissibility attenuation with the stiffness and damping values of the rubber mounts.

2. Methodology

In this analysis a single cylinder four stroke engine with 70 cc engine displacement is used. The dynamic characteristics of the handlebar is important and can be obtained from the experimental modal analysis (He and Fu 2001). The dimensions of the motorcycle handlebar are measured and modelled in the LMS Test Lab using the geometry workbench. Figure 1(a) shows measurement nodes located on the motorcycle stock and suspended handlebars labelled with stickers. Modal analysis was carried for the motorcycle handlebar mounted on the motorcycle structure. A small light weight accelerometer (Dytran 3055B2T) was used to measure the vibration response while an impact hammer (Krisler 9724A5000) excite the motorcycle handlebar. Both are connected to the 8-channel LMS SCADAS mobile. The accelerometers are calibrated using the calibration exciter type 4294 Brüel & Kjær.

Transmissibility measurement of both stock and suspended handlebar is carried out to quantify the attenuation level. Two accelerometers located at the input and output nodes measure the respective acceleration. The first accelerometer is placed on the motorcycle chassis and the second accelerometer is placed on the handlebar adjacent to the throttle unit.

Figure 1(a) and 1(b) shows the experimental setup for the in laboratory and on the road vibration analysis. The measurement instruments include two accelerometers (Dytran 3055B2T) for the z and y axis on the rider right hand, a tachometer (OPTEL-THEVON 152 G7 5V GPK XSI), calibrator (B&K 4294), data acquisition software (LMS Test Express), data acquisition hardware (LMS Scadas Mobile) and a portable work station. The measurement is carried out on straight-flat asphalt road surface with a period of 20 seconds data acquisition time. Accelerometers are located on the rider right hand to measure the hand vibration. The tachometer is mounted on the engine chassis of the motorcycle to measure the speed of the engine. The sensors are connected to the LMS Scadas Mobile running on LMS Test Express and a portable computer for the data acquisition system.

The modal model of the suspended handlebar used in the structural dynamic modification (SDM) within the LMS Test Lab software is used for the evaluation of the effect of stiffness and damping values of the rubber mounts on the dynamic characteristic of the suspended handlebar. Figure 1 (a), shows four rubber mounts installed on the suspend the handlebar. The existing value of the stiffness and damping in the z-axis are 600 kNm^{-1} and 6 kgs^{-1} and for the y-axis are 150 kNm^{-1} and 6 kgs^{-1} respectively. The optimization search covers the 0% - 300% of existing stiffness and damping values with the minimization of FRF area performance index (V.E.T. Index).

IOP Conf. Series: Materials Science and Engineering **815** (2020) 012013 doi:10.1088/1757-899X/815/1/012013

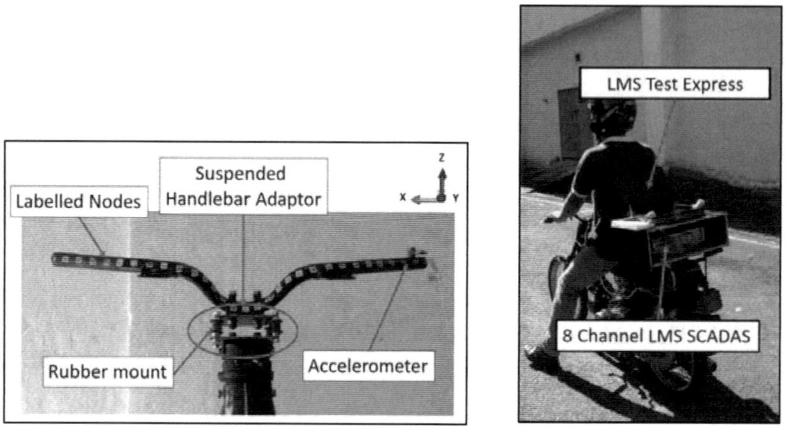

(a) Modal analysis setup (b) On-road vibration

Figure 1. Experimental analysis setup

3. Results and discussions

3.1. Modal analysis

3.1.1. Modal analysis of stock handlebar

Figure 2 shows the stock handlebar FRF stabilization curve in the y and z axis. The first five modes extracted below 200 Hz are at 31.92 Hz, 35.60 Hz, 45.12 Hz, 72.17 Hz and 152.62 Hz for the y-axis and four modes at 46.05 Hz, 56.11 Hz, 64.54 Hz and 156.70 Hz for the z-axis.

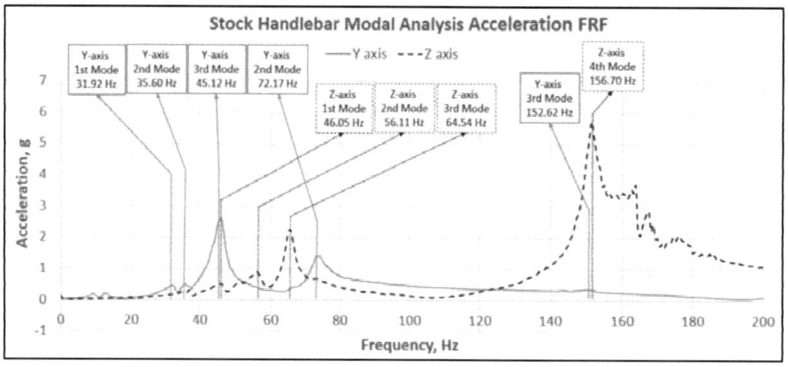

Figure 2. Stock Handlebar y- and z-axis FRF curve

3.1.2. Modal analysis of suspended handlebar

Based on figure 3, for the z-axis, the six modes extracted from the FRF curve on the z-axis at 27.00 Hz, 38.90 Hz, 51.02 Hz, 69.13 Hz, 95.11 Hz and 121.60 Hz. All of the modes extracted is selected to form the modal model of the respective axes for the SDM.

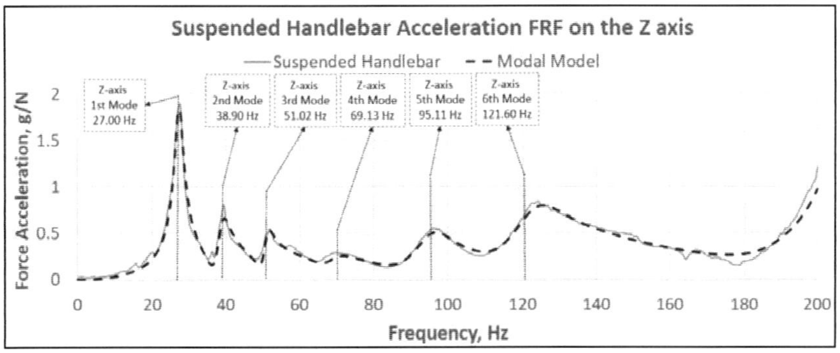

Figure 3. z-axis Suspended Handlebar and Modal Model FRF curve

3.2. Transmissibility measurement

Figure 4 shows the acceleration transmissibility ratio of the stock and suspended handlebar for the z-axis at the frequency range from 0 - 200 Hz. Based on figure 4, the z-axis transmissibility of the stock handlebar is in general higher than the suspended handlebar except at frequency of 25 Hz and 70 Hz. The highest transmissibility of the stock handlebar is 4.0 at 4 Hz while the suspended handlebar is 3.8 at 25 Hz. Both of the axes show an acceleration transmissibility reduction that dependent on the frequencies which most of the reductions are on the motorcycle typical cruising frequency range of 50 – 58 Hz or at the engine speed of 3000 – 3500 rpm. From this estimation, the suspended handlebar will reduce the vibration response induced on the rider during operating conditions.

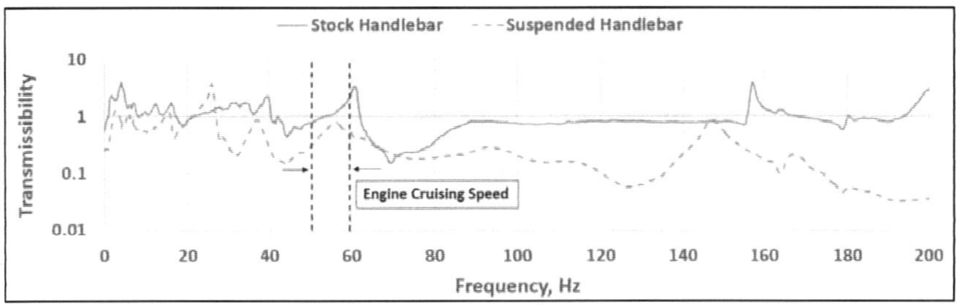

Figure 4. Motorcycle handlebar z-axis transmissibility

3.3. Hand-arm vibration

Based on the biodynamic axes of hand model, the measurement on the accelerometers mounted on the rider hand are used. For the z-axis response in figure 5 (a), the stock handlebar vibration displays two cluster of peaks, for the lower speed range of 1500 – 2750 rpm the clustering frequencies is at 46.05 Hz which coincide with the first mode of the z-axis. At the speed range of 3000 – 4000 rpm, the peaks clustered at the frequency of 64.54 Hz which coincide with the third mode of the z-axis. This speed independent constant frequency response indicate that the vibration is resonance- related. While during interchanging of clustering peaks, during the speed of 2750 – 3000 rpm the peaks are significantly clustered at the second mode frequency of about 56.11 Hz. The speed dependent and frequency dependent nature of the peaks can be attribute to the engine unbalance vibration. It is clear, that the stock handlebar suffered from the large amplitude vibration of the natural frequencies. For the suspended handlebar response on the z-axis in figure 5 (b), the installation of suspended handle altered the dynamic characteristic of the handlebar. The peaks clustered around the 46.05 Hz, 56.11 Hz and 64.54 Hz frequencies. With the introduction of the rubber mounts in the suspended handlebar the peaks pattern from the speed range of 1500 – 4000 rpm still exist. However, the peaks are effectively reduced in vibration amplitude for the suspended handlebar case especially in the speed range 3000 – 3750 rpm which are significantly important as it covers the cruising speed of the motorcycle at the speed of 50 – 60 kmh^{-1}. The data for the on the road test, verifies the effectiveness of the suspended handlebar in reducing hand arm vibration.

In order to see it more clearly, the difference in response of the stock and suspended handlebar at a single typical engine cruising speed is selected. Figure 6 shows the weighted FRF acceleration of the z-axis for the cruising engine speed of 3000 rpm at 60 kmh^{-1}. The peak weighted acceleration of the stock handlebar is 1.33 ms^{-2} at 56.25 Hz and after the installation of suspended handlebar, the highest peak is 0.52 ms^{-2} at 52.50 Hz indicating a reduction of 61 %. For, the FRF area performance index is 25.54 $kgms^{-3}$ for the stock handle and 13.05 $kgms^{-3}$ for the suspended handle gives a vibration reduction of 49 %. It is shown that the y-axis is more prone to road-tire vibration compare to z-axis as the y-axis exhibit a higher or equal magnitude of vibration acceleration on the lower frequency spectrum compared to higher frequency spectrum.

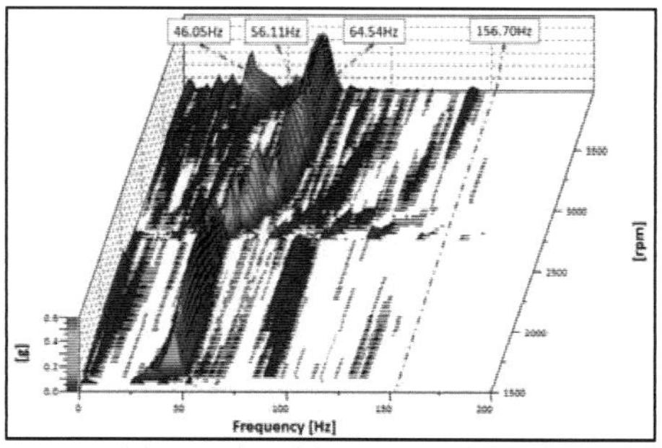

(a) z-axis with stock handlebar

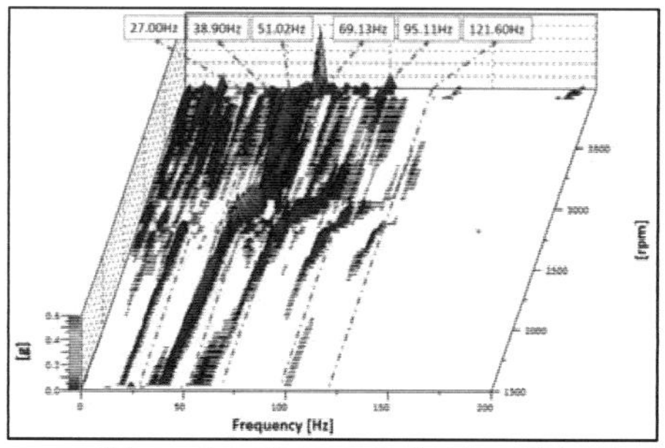

(b) z-axis with suspended handlebar

Figure 5. Hand acceleration spectra waterfall graph (on the road z-axis)

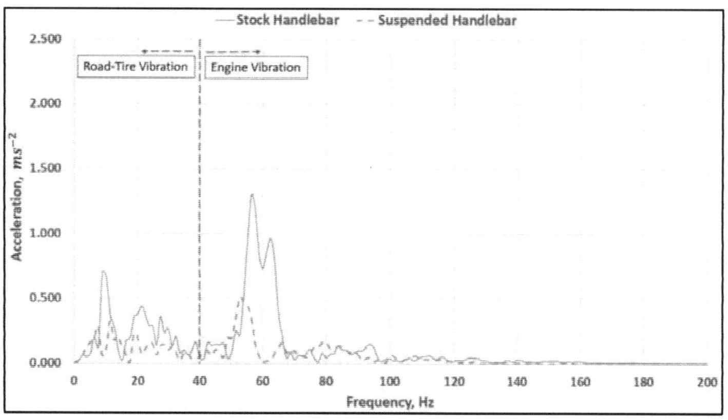

Figure 6. Weighted FRF rms acceleration graph of the hand at engine speed of 3000 rpm. (z-axis)

Figure 7 shows the more detail hand-arm vibration values a_{hv} for the stock and suspended handlebar and its reduction percentage at engine cruising speed range of 3000 – 3500 rpm (60-75 kmh^{-1}). It is shown that the vibration values of the stock handlebar are higher than suspended handlebar within the speed range. The peak vibration value of the stock handlebar is 16.51 ms^{-2} at 3375 rpm and the suspended handlebar vibration values of 3.41 ms^{-2} with the reduction of 79 %. Based on the hand-arm vibration exposure, the existing stock handlebar has a limit of 11 and 44 minutes of exposure before reaching an exposure action value (EAV) and exposure limit value (ELV) respectively. While the suspended handlebar has a 258 minutes and 1032 minutes of exposure period before exceeding the (EAV) and (ELV) respectively. The installation of suspended handlebar has reduced the hand-arm vibration exposure point per hour from 545 to 23 and dramatically increase the time required to reach the exposure limits (EAV) and (ELV) by 20 times.

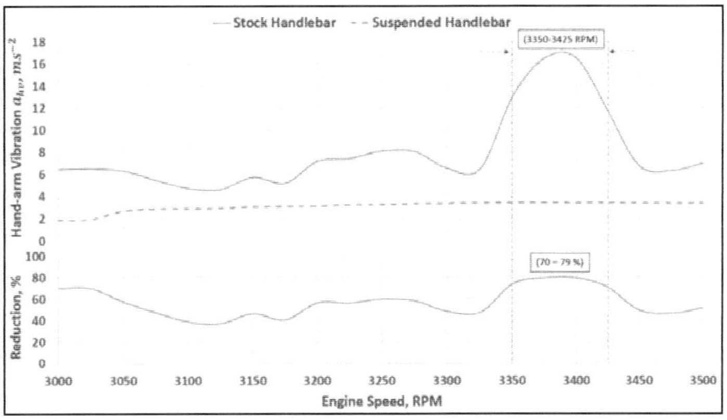

Figure 7. Hand-arm vibration acceleration values and percentage of reduction against cruising engine speed.

3.4. Optimization of rubber mount properties

Referring to figure 3, from the modal analysis of the suspended handlebar, the validated modal model of both axes are used in the structural dynamic modification. Figure 8 shows the suspended handlebar optimized value of rms V.E.T. Index under a structural dynamic modification of rubber mount stiffness and damping properties. The baseline rms V.E.T. index calculated is 5.782 $kgms^{-3}$ at the total calculated stiffness and damping values at 150.0 kNm^{-1} and 6.0 kgs^{-1} in the y-axis and 600.0 kNm^{-1} and 6.0 kgs^{-1} in the z-axis. With the optimization of rms V.E.T. Index toward the lowest index value, the measured index is 0.164 $kgms^{-3}$ at the addition of stiffness and damping from baseline value of 300 % and 90% respectively. The total calculated stiffness and damping values of the optimized V.E.T. index is at 472.5 kNm^{-1} and 11.4 kgs^{-1} in the y-axis and 2400.0 kNm^{-1} and 18.0 kgs^{-1} in the z-axis.

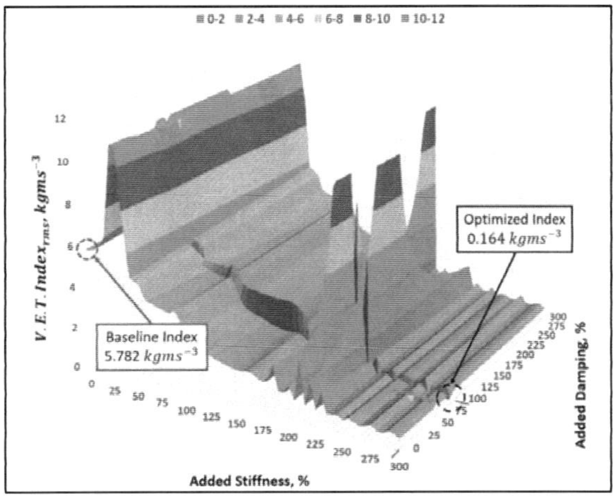

Figure 8. Optimization of suspended handlebar

4. Conclusion

A suspended handle adaptor has been being developed to attenuate the motorcycle handlebar vibration transmitted to the rider. The highest attenuation of the hand-arm weighted RMS acceleration is 79 % based on the on-road test which gives a longer vibration exposure time by 20 times. Based on the optimization of rubber mount properties of the suspended handlebar in SDM, the lowest V.E.T. index calculated is 0.164 $kgms^{-3}$ from the baseline index of 5.782 $kgms^{-3}$. The total calculated stiffness and damping value is at 472.5 kNm^{-1} and 11.4 kgs^{-1} in the y-axis and 2400.0 kNm^{-1} and 18.0 kgs^{-1} in the z-axis.

5. References

[1] I. S. O. ISO, "5349-1: Mechanical vibration-measurement and evaluation of human exposure to hand-transmitted vibration-part 1: general requirements," *Geneva, Switz. Int. Organ. Stand.*, 2001.

[2] I. ISO, "5349-2: Mechanical vibration- Measurement and evaluation of human exposure to hand-transmitted vibration-part 2: Practical guidance for measurement at the workplace," *Geneva, Switz. Int. Organ. Stand.*, 2002.

[3] S. M. Mirbod, H. Yoshida, M. Jamali, K. Masamura, R. Inaba, and H. Iwata, "Assessment of hand-arm vibration exposure among traffic police motorcyclists," *Int. Arch. Occup. Environ. Health*, vol. 70, no. 1, pp. 22–28, 1997.

[4] L. M. Roseiro, M. A. Neto, A. M. Amaro, C. J. Alcobia, and M. F. Paulino, "Hand-arm and whole-body vibrations induced in cross motorcycle and bicycle drivers," *Int. J. Ind. Ergon.*, vol. 56, pp. 150–160, 2016.

[5] M. Yokomori, S. Yamada, T. Nakagawa, and T. Matsumoto, "The Vibration of the Handle-Bars of a Motorcycle in Running on the Paved Road," *Japanese J. Ind. Heal.*, vol. 23, no. 2, pp. 134–40, 1981.

[6] K. Karmegam *et al.*, "A study on motorcyclist's riding discomfort in Malaysia," *Eng. e-Transaction*, vol. 4, no. 1, pp. 39–46, 2009.

[7] Jeyapal and B. R, "Judging the discomfort of vibration on motorbike," *Inst. Sound Vib. Res. Univ. Southampton, UK.*, 2014.

[8] N. K. Khamis, B. Deros, F. R. Mismail, and N. H. M. Tahir, "Motorcycle deliveryman's perceptions on riding conditions," *Malaysian J. Public Heal. Med.*, vol. 1, no. Specialissue1, pp. 1–5, 2016.

[9] M. I. Mohd Hafzi, S. Rohayu, P. Noor Faradila, and S. V Wong, "Prevalence and risk factors of musculoskeletal disorders of motorcyclists," *Malaysian J. Ergon.*, vol. 1, pp. 1–10, 2011.

Acknowledgement

The authors would like to thank the Ministry of Higher Education and USM for providing the facilities and the research opportunity through Bridging Grant A/C 304/PMekanik/6316214

IOP Conf. Series: Materials Science and Engineering **815** (2020) 012014 doi:10.1088/1757-899X/815/1/012014

Development of an Ultrasonic Scalpel

Yi Bin Ngo, Zaidi Mohd Ripin*, Chan Ping Yi, Muhammad Ikhwan Zaini Ridzwan, Wan Mohd Amri Wan Mamat Ali and Baharum Awang

TheVibrationLab, School of Mechanical Engineering, Universiti Sains Malaysia, 14300 Nibong Tebal, SPS, Pulau Pinang, Malaysia.

*Corresponding author: mezaidi@usm.my

Abstract. This paper explains the development of an ultrasonic scalpel for soft tissue dissection. It is used in both open and laparoscopic surgery to coagulate and dissect the soft tissue. The ultrasonic scalpel has been designed, prototyped, and characterized. The development includes its power supply with ultrasonic-wave-signal generator, ultrasonic transducer, ultrasonic amplifier, a pair of end effectors, and a casing with trigger mechanism. Conceptual design was done by using SolidWorks, and the conceptual design was simulated and refined until optimum design is obtained. The prototype of the final design was done by using machining process for the metal components and rapid prototyping for the plastic parts which are finally assembled for characterization purpose, which includes the measurement of natural frequency, vibration stroke, and its capability to dissect soft tissue. The result shows that the natural frequency of the ultrasonic scalpel is 29.3 kHz, with peak to peak vibration stroke of $40\mu m$. The ultrasonic scalpel can dissect muscle tissue of chicken in $16.6s$ at the maximum temperature along the dissection site of 81.7°C.

1. Introduction

As early as 1993, ultrasonic cautery was developed by Amaral, called 'laparoscopic scalpel' for laparoscopic surgery by allowing simultaneous coagulation and dissection of blood vessel [1]. Laparoscopic surgery is the first type of minimally invasive surgery where it is a surgery done through one or more small incisions, rather than a larger incision through abdominal wall. It is done by using small tubes, imaging system, and surgical instruments with extension rod [2]. Energy devices of laparoscopic instruments such as monopolar, bipolar, and ultrasonic scalpel are widely used during the operation [3]. As compared to the electrosurgery devices, ultrasonic scalpel reduces the operating time, difficulty, blood loss, and lateral thermal spread [4], [5].

Within the ultrasonic scalpel, alternating current is applied to electric generator which convert to direct current and into oscillated current at desired frequency. Piezoceramic acts as transducer to converts electrical energy into mechanical vibration energy at frequencies ranges from 20kHz to 60kHz. The vibration motion oscillates the active blade of the ultrasonic scalpel linearly and the active blades movement ranges from 50 to 100μm [6]. Tissue is held between the active and counter blade and heat is generated from the friction between active blade and the counter blade to denature the protein in the tissue resulting in coagulation. The advantages of ultrasonic scalpel include less instrument traffic as it can perform vessel sealing and tissue cutting simultaneously, and low smoke generation [7]. The temperature generated ranges from 50°C to 300°C, which reduces the lateral thermal spread and less charring [8]. In the case of lower density tissue with high water content, the intracellular water is vaporized by the heat generated causing 'cavitation effect' further assisting in the dissection by separating tissue layers [3]. The efficiency of dissection by ultrasonic scalpel is dependent on the nature of the tissue such as its water content, and the tissue strength which is directly related to the type of tissue, amount of collagen and its organization.

When applying ultrasonic scalpel to a tissue, there are three main effects on the tissue which are cavitation, coagulation, and cutting. Firstly, in the cavitation effect, oscillating pressure field is produced due to the linear stroke from the active blade and causing the expansion and contraction of intracellular fluid and the tissue around the dissected site at ultrasonic frequency. When it is in the expansion stage, the pressure of the intracellular fluid dropped below its vapor pressure, causing vapor bubbles to be generated within the tissue. In contraction stage, the pressure is then raised rapidly causing the vapor bubbles to collapse. Since this happened is at ultrasonic frequency, shock wave is produced, and jet-like ejections is generated into the intracellular fluid. These phenomena happened in micro-second scale and over a period, the tissue will be dissected [9]. Cavitation is essential because it causes the separation of tissue planes facilitating the dissection process later. Next, for the coagulation, it occurs when the frictional heat generated between tissue and active blade is transmitted into tissue, which might heat up to 60°C to 100°C. Denaturation of protein or collagen starts and result in occlusion of blood vessel. Denaturation of protein happened because the tertiary hydrogen bonds between collagen and protein is broken. The proteins denature and convert from colloidal proteins into an insoluble gel that helps on occlusion. Lastly, the cutting effect is achieved by the high frequency oscillation of the active blade applied to the tissue [10].

The project has two objectives highlighted below:

1. To develop an ultrasonic scalpel as it can be used for cutting and cauterizing tissue. The design of ultrasonic scalpel includes its ultrasonic transducer, ultrasonic horn, trigger mechanism, active and counter blades, and the casing.
2. To characterize the designed ultrasonic scalpel in terms of frequency, stroke, elevated temperature, and the ability to cut tissues.

2. Methodology

2.1. Conceptual Design of Ultrasonic Scalpel

The conceptual design of ultrasonic scalpel done in the SolidWorks includes two independent working systems, which is as shown in figure 1.

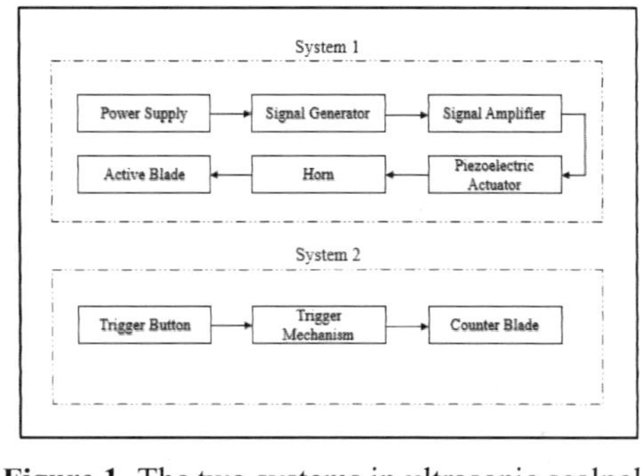

Figure 1. The two systems in ultrasonic scalpel.

In the System 1, oscillated electric signal from the signal generator is amplified to actuate the piezoceramic. Through the inverse piezoelectric effect, the oscillated current is converted into vibration stroke. With the stroke amplifier, the active blade will vibrate at a desired range of stroke. In the System 2, the mechanism is functioned to transmit the load applied by the surgeon, to the counter blade. In ideal condition, there is no energy loss due to damping of the system

IOP Conf. Series: Materials Science and Engineering **815** (2020) 012014 doi:10.1088/1757-899X/815/1/012014

within the mechanism, the amount of energy of surgeon applied should be all goes to the counter blade to hold the soft tissue firmly.

2.2. Finite Element Analysis
After the conceptual design is done, finite element analysis is carried out to simulate the natural frequency of the system. The result obtained from simulation is used to validate the result obtained from the measurement of natural frequency of the system. SolidWorks software is used to simulate the system.

2.3. Prototyping
In the prototyping stage, the rapid prototyping technology was be used for plastic parts while the machining process was used for metal parts. FDM 3D Printer, CNC, EDM Wire Cut, and some other machining processes have been applied to fabricate the prototype.

2.4. Characterization of Ultrasonic Scalpel
After the prototype of designed ultrasonic scalpel was created, it was characterized to describe about its key features in soft tissue dissection, which are natural frequency of the system, vibration stroke produced by ultrasonic scalpel, time taken for soft tissue dissection, and maximum temperature elevated during dissection.

2.5. Measurement of Natural Frequency of the Ultrasonic Scalpel
To measure the natural frequency of the system, signal generator, oscilloscope, and the ultrasonic transducer can be connected through electrical circuit, as shown in the figure 2.

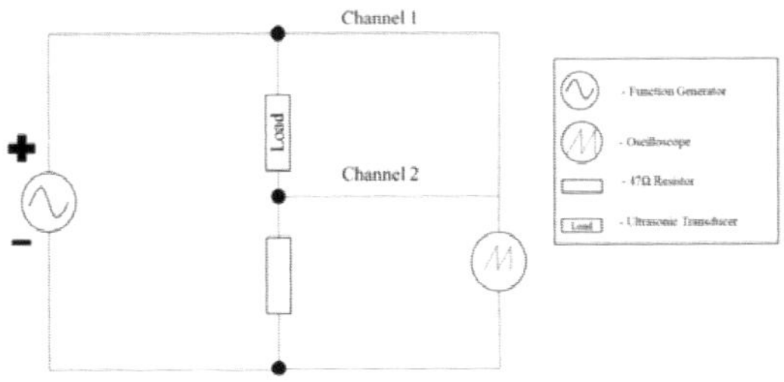

Figure 2. Circuit design for measurement of the natural frequency

When the ultrasonic scalpel is excited at its natural frequency, the impedance of the EUA, Z_{load} is at its lowest point as the ultrasonic scalpel vibrates at its maximum amplitude. The 47Ω resistor is connected in series to the ultrasonic scalpel, both loads share the same amount of current within the circuit. Since $Z_{load} \ll 47\Omega$, the voltage drops across the ultrasonic scalpel is at the least amount. Most of the voltage drops across the 47Ω resistor and it shows the maximum potential difference across the circuit. If ultrasonic scalpel vibrates at its natural frequency, the difference between the root mean square voltage, V_{rms} should be at its minimum, and the two voltage signals across ultrasonic scalpel and resistor should be ideally in phase.

2.6. Measurement of Vibration Stroke of the Ultrasonic Scalpel
Next, to measure the vibration stroke produced by the ultrasonic scalpel, electrical circuit connection, as shown in the figure 3 will be followed.

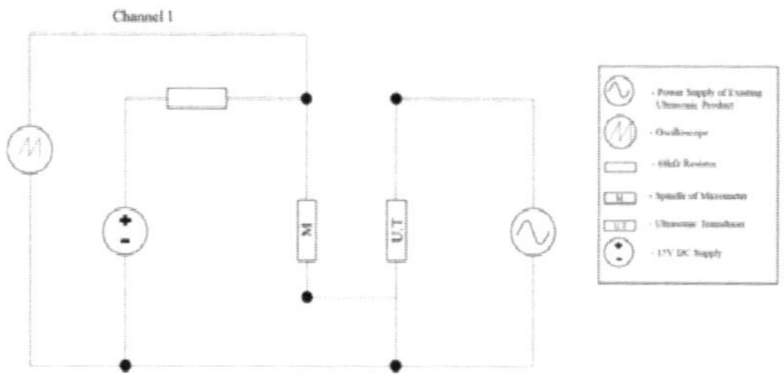

Figure 3. Circuit for stroke measurement of the ultrasonic scalpel.

From the schematic diagram as shown in figure 3, it can be observed that the circuit on the left is open circuit, no current flows across the 68kΩ resistor and micrometre, hence there is no voltage drops across the two components. The oscilloscope that connected to measure the voltage signal of micrometre will shows the 15V of power supply on screen with zero current. However, when the spindle of micrometre is in contact with the active blade of US, the micrometre is then grounded through the US. The DC power supply provide 15V of voltage and flow across the 68kΩ resistor and micrometre. Since micrometre and resistor are connected in series, the current across the two components is constant. By applying the Ohm's Law, voltage drop is higher at the components with higher resistance. Since the spindle of micrometre is a good electrical conductor, it has negligible resistance as compared to 68kΩ resistor, thus voltage drops across the micrometre is approximately zero. The oscilloscope will show a zero-voltage signal when the spindle of micrometre and active blade of US is in contact.

2.7. Dissection of Soft Tissue by Ultrasonic Scalpel
In this section, the fabricated ultrasonic scalpel was used to test for soft tissue dissection. In this experiment, the time taken for soft tissue dissection and the maximum temperature variation at the dissection site were measured. In this experiment, ultrasonic scalpel will be handheld to dissect the soft tissue. The functionality of the trigger mechanism was tested on the ultrasonic scalpel. In the sample preparation, chicken breast was chosen as sample and sliced into small piece with its width is controlled within 5mm.

3. Results and discussions
3.1. Conceptual Design of Ultrasonic Scalpel
Figure 4 depicts the prototype of the ultrasonic scalpel with the labelled components. In the System 1, the oscillated electric current through power supply actuated the piezoceramics. The vibration stroke produced is then amplified through the horn and transmitted to the active blade to cut the soft tissue (the top blade as in figure 4). The active blade is driven by the ultrasonic motor which is cylindrical and located in the motor housing which is the bulky cylinder in the bottom half of the scalpel. For the System 2, the load applied using a scissor like mechanism. The counter blade is flat and together with the active blade, shear the soft tissue located in between the two blades for dissection.

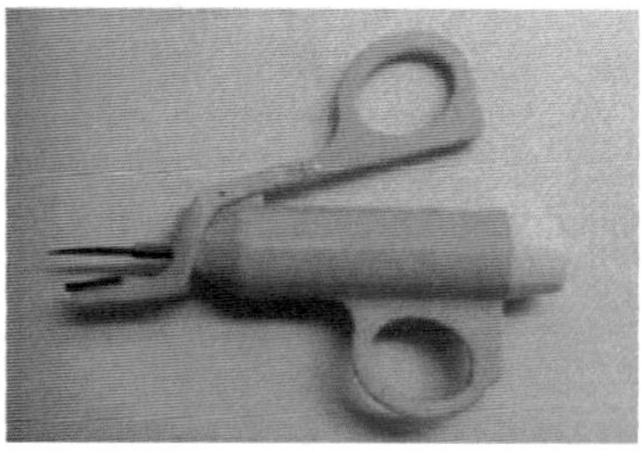

Figure 4. The prototype ultrasonic scalpel.

3.2. Simulation and Measurement of Natural Frequency of Ultrasonic Scalpel
The results obtained from the simulation and measurement are listed in table 1.

Table 1. Results of simulation and measurement.

	Measured Natural Frequency, $f_{c,exp}$ (kHz)	Simulated Natural Frequency, $f_{c,s}$ (kHz)	Percentage Difference, %
Without Active Blade	43.5	41.8	3.82
With Active Blade	29.3	28.2	3.53

The result shows that the first natural frequency of the desired mode shape is, $\boldsymbol{f_{c1,2} = 28.2\ kHz}$. As observed, the natural frequency is dropped from 41.84 kHz to 28.245 kHz. The natural frequency of a system is inversely proportional to the mass. When active blade is added to the system, the active blade contributes additional mass to the system. With the increase in mass of system, the natural frequency of system is dropped.

3.3. Measurement of vibration stroke of the ultrasonic scalpel
The results obtained are listed in table 2. Figure 5 shows the example of voltage signal obtained when the spindle of micrometre is at the peak of vibration stroke.

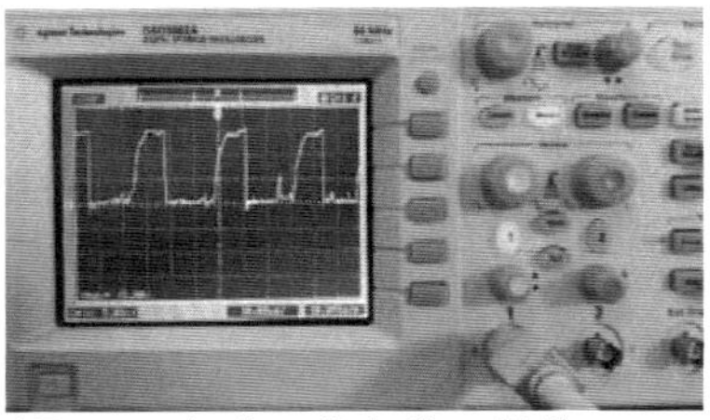

Figure 5. Voltage Signal Detected when Spindle of Micrometre at Peak Position.

Table 2. Vibration Stroke of Ultrasonic Scalpel.

Trial	Reading on Micrometer at Peak Position, mm	Reading on Micrometer at Rest Position, mm	Forward Stroke, μm	Peak to Peak Vibration Stroke, μm
1st	0.45	0.47	20	40
2nd	0.03	0.05	20	40
3rd	0.37	0.35	20	40

From the result obtained, it shows that the peak to peak vibration stroke of System 1 in ultrasonic scalpel is $40\mu m$ throughout the 3 trials. The type of micrometre used is depth micrometre with resolution of $\pm 0.01mm$. Thus, the tolerance of for the peak to peak vibration stroke is $\pm 0.005mm$.

3.4. Dissection of Soft Tissue by Ultrasonic Scalpel
Figure 6 shows the sample tissue after dissection by using prototype of ultrasonic scalpel. The time taken for dissection and maximum temperature variation during the dissection process are plotted in figure 7 and figure 8 respectively.

IOP Conf. Series: Materials Science and Engineering **815** (2020) 012014 doi:10.1088/1757-899X/815/1/012014

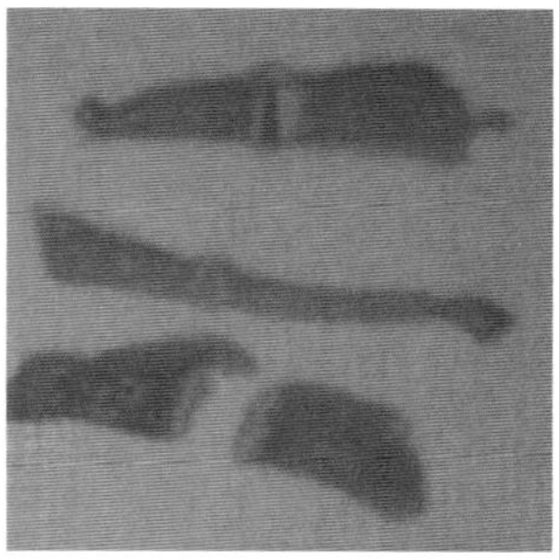

Figure 6. Tissues from the dissection.

In figure 7, the average time taken for the dissection by ultrasonic scalpel is 16.6s. In Fig.11, it shows that the average maximum temperature elevated at dissection site is 81.72°C.

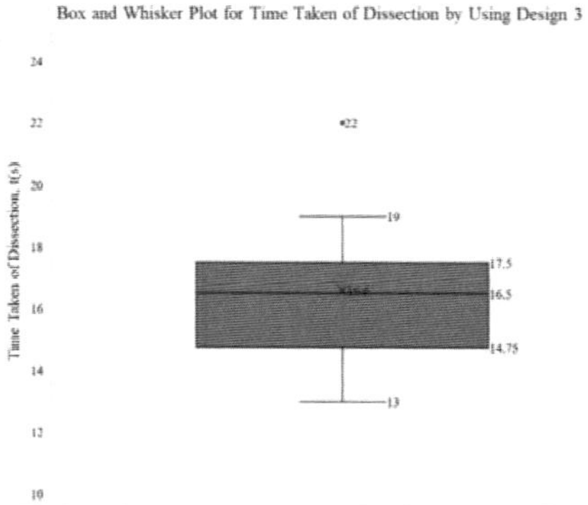

Figure 7. Time Taken for Dissection of Muscle Tissue by Ultrasonic Scalpel.

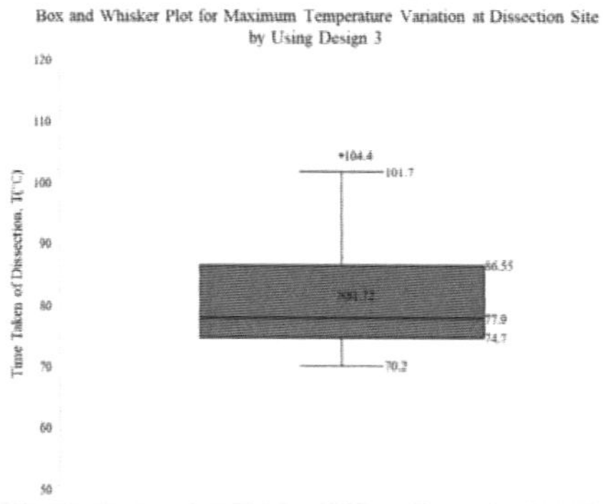

Figure 8. Maximum Temperature Variation at Dissection Site

One of the challenges in the development of energy scalpel is the collateral tissue damage. In this case because of the high internal temperature generated at the cut location there will also lateral thermal spread on the dissected tissue with the width of approximately 1mm from the dissection site on one side of dissection site. Thus, there is total lateral thread spread of less than 2mm on the sample of tissue. This is the proof where the frictional heat generated is enough to denature the protein within soft tissue. There is minimum amount of excessive heat generated to transmit into lateral thermal spread.

In the dissection process, tissue is held firmly with a certain amount force on it. Once the ultrasonic transducer is excited, the active blade on the soft tissue will start to oscillate back and forth at ultrasonic frequency. This will create a sliding effect on the tissue, and friction due to the sliding is generated at the dissection site. Friction has been continuously occurring between the surface of active blade and the soft tissue. Thus, frictional heat is generated on the dissection site. As dissection process is carried on, more heat is generated and causes the heat zone increased. The heat zone can be observed with the colour changed on the soft tissue from light pink colour to the white colour. This is happened when the protein within the region undergo denaturation. In addition, mist is released from the dissection site during the dissection process. This has proven that the vapor bubble created during the expansion stage from the intracellular fluid is collapsed. The thin film on the vapor bubble is spread out as the mist when the bubble is collapsed.

4. Conclusion

The ultrasonic scalpel has been developed and characterized. The natural frequency of the system is 29.3 kHz and capable of stroke of $40\mu m$. During the dissection, the time taken for dissection is as short as 16.6s while the maximum temperature elevated at the dissection site is 81.72°C.

5. References

[1] S. D. Lyons and K. S. K. Law, "Laparoscopic Vessel Sealing Technologies," *J. Minim. Invasive Gynecol.*, vol. 20, no. 3, pp. 301–307, 2013.
[2] R. Baxter, N. Hastings, A. Law, and E. J. . Glass, "Instruments and devices used in laparoscopic surgery," *Anim. Genet.*, vol. 39, no. 5, pp. 561–563, 2008.
[3] Y. Yadav, R. O'Shea, and F. Behnia-Willison, "Laparoscopic surgical tools: a review," *O&G Magazine*, p. Vol. 17 No 4, 2015.
[4] A. Upadhyay, A. K. Gupta, A. Karigoudar, N. Gupta, and U. Krishnegowda, "A Comparative Study Between Ultrasonic Dissector Versus Conventional Methods in Achieving Haemostasis," pp. 410–414, 2016.
[5] L. J. Hefermehl, R. A. Largo, T. Hermanns, C. Poyet, T. Sulser, and D. Eberli, "Lateral temperature spread of monopolar, bipolar and ultrasonic instruments for robot-assisted laparoscopic surgery," *BJU Int.*, vol. 114, no. 2, pp. 245–252, 2014.
[6] D. K. Dutta and I. Dutta, "The Harmonic Scalpel," *J. Obstet. Gynecol. India*, vol. 66, no. 3, pp. 209–210, 2016.
[7] S. Lyons and D. N. Kee, "Laparoscopic energy sources," *O&G Magazine*, p. Vol. 17 No 4, 2015.
[8] F. J. Kim *et al.*, "Temperature safety profile of laparoscopic devices: Harmonic ACE (ACE), Ligasure V (LV), and plasma trisector (PT)," *Surg. Endosc. Other Interv. Tech.*, vol. 22, no. 6, pp. 1464–1469, 2008.
[9] R. Lockhart *et al.*, "Silicon micromachined ultrasonic scalpel for the dissection and coagulation of tissue," *Biomed. Microdevices*, vol. 17, no. 4, pp. 1–12, 2015.
[10] W. Sasi, "Dissection by Ultrasonic Energy Versus Monopolar Electrosurgical Energy in Laparoscopic Cholecystectomy," *JSLS J. Soc. Laparoendosc. Surg.*, vol. 14, no. 1, pp. 23–34, 2010.

Acknowledgement
The authors wish to thank the Universiti Sains Malaysia for the RUI grant account number RUI account no: 1001/PMEKANIK/8014070.

Modelling of the Human Body Under the Exposure to Recumbent Whole-body Vibration

L K Lucas[1] and Y H Ko[1]*

[1]Department of Mechanical Engineering, Faculty of Engineering and Technology (FOET), Tunku Abdul Rahman University College (TARUC).

*Corresponding author: koyh@tarc.edu.my

Abstract. Human is exposed to whole-body vibration either in sitting, standing or recumbent position. Many studies had been carried out to study the vibration responses and develop the mathematical modelling in sitting and standing vertical position. According to previous studies, the resonant frequencies of the human body were identified to be in the range of 4-6 Hz and 8-12 Hz in the vertical position. Recumbent whole-body vibration also plays an important role as many situations in our everyday life are related to recumbent whole-body vibrations such as riding a train, ambulance transporting a patient or even sleeping. However, the studies of recumbent whole-body vibration are very lacking. Therefore, the main objective of this study is to develop a mathematical model to identify the response when a human body is subjected to recumbent whole-body vibrations. In this study, a nine degree of freedom mathematical model was developed, and the responses of the body were analyzed using Matlab Simulink. The overall resonant frequency of the body was found to be 1.4 Hz. Meanwhile, the head and neck had two resonant frequencies, which were 1.3 and 3.28 Hz for the head, and 1.4 and 3.4 Hz for the neck. This newly developed mathematical model can be used as a fundamental model to predict the response of any recumbent whole-body vibration.

1. Background

Nowadays, human exposure to vibration is quite common whether in their occupational life or even at home. This is especially the case for jobs like construction workers, drivers for heavy vehicles such as trucks and busses. From previous works, human workers exposed to hand-transmitted vibrations (HTV) have risks of reduction in grip strength, manual dexterity and sensory perception. [1]

Based on previous works, it was concluded that whole-body vibrations (WBV) can be very complicated when it comes to human responses. It is also clear that prolong period of WBV exposure may cause several unwanted effects such as damaging health or cause discomfort which disrupts daily activities. [2] When exposed to vibrations, the body endures stresses which have a mechanical effect on it depending on the dynamic properties of the body. By understanding all these effects and harm vibrations can cause to the human body, it is clear that human body vibrations should be further studied and look into to provide an economical method to deal with these issues.

1.1. Whole-Body Vibration in Recumbent Position

The human body can be exposed to vibrations in three main different postures which are sitting, standing and the recumbent position. Vibrations can cause discomfort or can even threaten human health. For the recumbent position, discomfort from WBV is relevant when it comes to transportation where the passenger is lying down, such as ambulances and trains, or even houses where it is near to railway tracks and airports. Despite the many cases of recumbent WBV, the number of studies of this area is relatively small compared to those of in sitting and standing posture. [3]

However, the effects of vibration could also be made to good use. A study showed that vibration can also be used to induce sleep with the right parameters. Previous studies had concluded that babies sleep well when being rocked gently as this rhythmic motion reassembles the movement and feeling while in the mother's womb. With deeper understanding of the human response under WBV, it would be possible to reproduce this motion using machines to put a person, such as an insomnia patient or infant to sleep without harm. [4]

1.2. Research Objectives

The main objective of this research is to construct a mathematical model of the human body in the recumbent position, as to simulate and study the response of recumbent whole-body vibration. The results from the simulation are also analyzed to obtain the resonant frequency of the different body parts.

1.3. Problem Statement

In this study, WBV is studied in three different postures, which are sitting, standing and supine. Previous studies were mainly on vertical vibrations in sitting positions. The number of studies of recumbent WBV is extremely low compared to the other two postures mentioned above. This was because WBV started as a study to solve problems such as vehicle comfort and pilot ejection injuries. Since recumbent WBV can now occur in many situations such as sleeping in rocking motion or in the trains, inter-hospital ground transports, infant incubators, the study has become more relevant to the society's current problems.

2. Methodology

For this study, a lumped-parameter study of the recumbent WBV is conducted. A mathematical model of the recumbent human body is developed and simulated under WBV using Matlab Simulink.

2.1. Development of Mathematical Model

The model for the human body is adapted from [5]. The original model was developed to simulate vertical and horizontal vibrations on the seated human body. Hence, this model consists of vertical and horizontal stiffness data which will be adapted in our simulation. The mathematical model is modified accordingly to suit the study of recumbent WBV. The modified model consists of 9 masses, similar to the adapted model, excluding the lower and upper arms. The arms are excluded from the model as they play a less significant role in terms of safety and comfort. Additional parameters were added to the model to match the recumbent body. The head, pelvis and thoracic spine is connected to the vibrating platform which simulates the real-life recumbent situation.

The values of the thoracic spine connection were justified and adapted from [6]. Meanwhile, the head and pelvis parameters were adapted from [5], with the assumption of the head is supported by a pillow which the mechanical parameters were equal to the buttocks tissue. The complete data of mass and mechanical parameters of the model are shown in Table 1 and 2 respectively. The mathematical model was developed as shown in Figure 1.

IOP Conf. Series: Materials Science and Engineering **815** (2020) 012015 doi:10.1088/1757-899X/815/1/012015

Table 1. Values for Mass

Mass No.	Body Segments	Mass (kg)
M_0	Platform	-
M_1	Head	5.445
M_2	Cervical Spine	1.084
M_3	Thoracic Spine	4.806
M_4	Lumbar Spine	2.002
M_5	Torso	32.697
M_6	Thorax	13.626
M_7	Diaphragm	0.454
M_8	Abdomen	5.906
M_9	Pelvis	27.174

Table 2. Values for Mechanical Properties

Mass No.	Parameter Description	Stiffness (N/m)	Damping Coefficient (kg/s)
12	Head-Neck	5364	365.1
23	Neck-Thoracic	5364	365.1
34	Thoracic-Lumbar	5364	365.1
35	Thoracic-Torso	5364	365.1
36	Thoracic-Thorax	5364	365.1
49	Lumbar-Pelvis	5364	365.1
55	Torso-Thorax	5364	365.1
67	Thorax-Diaphragm	89.41	29.8
78	Diaphragm-Abdomen	89.41	29.8
89	Abdomen-Pelvis	89.41	29.8
90	Pelvis-Platform	2550	37.8
30	Thoracic-Platform	2300	154
10	Head-Platform	2550	37.8

The item numbering in Table 2 indicates the 2 segments involved in the parameter. For example, k_{12} indicates the stiffness connecting the head and the neck.

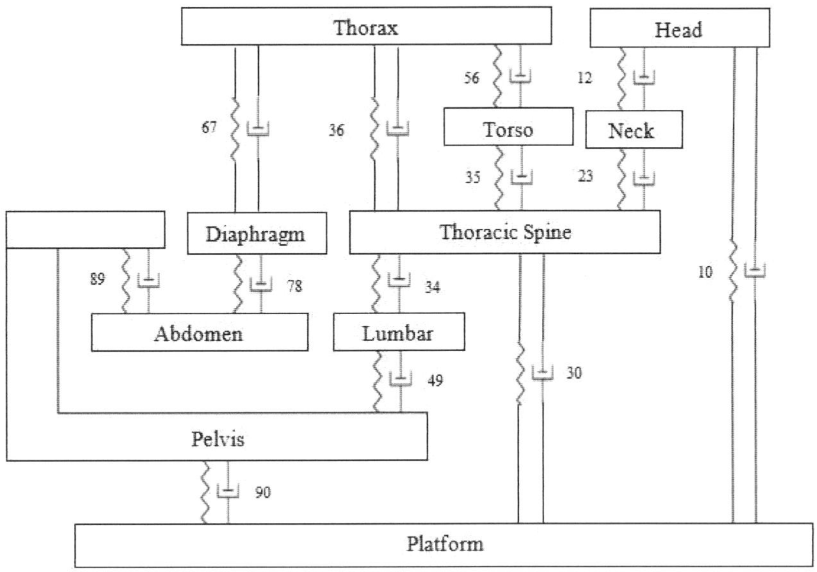

Figure 1. Mathematical model of recumbent whole-body vibratory system.

2.2. *Equations of Motion*

Equations of motion are derived for each mass (Table 2.1). The harmonic force F(t) exerted by the platform is represented by *F* in the equations. The derivations of the masses are as follows.

$$\ddot{x}_1 = \frac{1}{m_1}[-(k_{12}+k_{10})x_1 - (c_{12}+c_{10})\dot{x}_1 + k_{12}x_2 + c_{12}\dot{x}_2 + F] \tag{1}$$

$$\ddot{x}_2 = \frac{1}{m_2}[k_{12}x_1 - (k_{12}+k_{23})x_2 + c_{12}\dot{x}_1 - (c_{12}+c_{23})\dot{x}_2 + k_{23}x_3 + c_{23}\dot{x}_3] \tag{2}$$

$$\ddot{x}_3 = \frac{1}{m_3}\begin{bmatrix} -(k_{36}+k_{30}+k_{35}+k_{23}+k_{34})x_3 - (c_{36}+c_{30}+c_{35}+c_{23}+c_{34})\dot{x}_3 \\ +k_{36}x_6 + c_{36}\dot{x}_6 + k_{35}x_5 + c_{35}\dot{x}_5 + k_{23}x_2 + c_{23}\dot{x}_2 + k_{34}x_4 + c_{34}\dot{x}_4 + F \end{bmatrix} \tag{3}$$

$$\ddot{x}_4 = \frac{1}{m_4}[k_{34}x_3 + c_{34}\dot{x}_3 - (k_{34}+k_{49})x_4 - (c_{34}+c_{49})\dot{x}_4 + k_{49}x_9 + c_{49}\dot{x}_9] \tag{4}$$

$$\ddot{x}_5 = \frac{1}{m_5}[k_{56}x_6 - (k_{56}+k_{35})x_5 + c_{56}\dot{x}_6 - (c_{56}+c_{35})\dot{x}_5 + k_{35}x_3 + c_{35}\dot{x}_3] \tag{5}$$

$$\ddot{x}_6 = \frac{1}{m_6}\begin{bmatrix} -(k_{67}+k_{36}+k_{56})x_6 - (c_{67}+c_{36}+c_{56})\dot{x}_6 + k_{67}x_7 \\ +c_{67}\dot{x}_7 + k_{36}x_3 + c_{36}\dot{x}_3 + k_{56}x_5 + c_{56}\dot{x}_5 \end{bmatrix} \tag{6}$$

IOP Conf. Series: Materials Science and Engineering **815** (2020) 012015 doi:10.1088/1757-899X/815/1/012015

$$\ddot{x}_7 = \frac{1}{m_7}[k_{67}x_6 + c_{67}\dot{x}_6 - (k_{67} + k_{78})x_7 - (c_{67} + c_{78})\dot{x}_7 + k_{78}x_8 + c_{78}\dot{x}_8] \quad (7)$$

$$\ddot{x}_8 = \frac{1}{m_8}[k_{89}x_9 + c_{89}\dot{x}_9 - (k_{89} + k_{78})x_8 - (c_{89} + c_{78})\dot{x}_8 + k_{78}x_7 + c_{78}\dot{x}_7] \quad (8)$$

$$\ddot{x}_9 = \frac{1}{m_9}[-(k_{89} + k_{90})x_9 - (c_{89} + c_{90})\dot{x}_9 + k_{89}x_8 + c_{89}\dot{x}_8 + F] \quad (9)$$

2.3. *Simulink Model*

The Simulink model was constructed based on the EOM derived using equations 1 to 9. After the model has been constructed. The sinestream input was set to range from 0.1 to 20 Hz with 0.1 increment. This range was selected as this range of frequency was considered to be the everyday transport vibrational frequency. [3] Due to large overall Simulink model, only the model for the head (M1) is shown as follows.

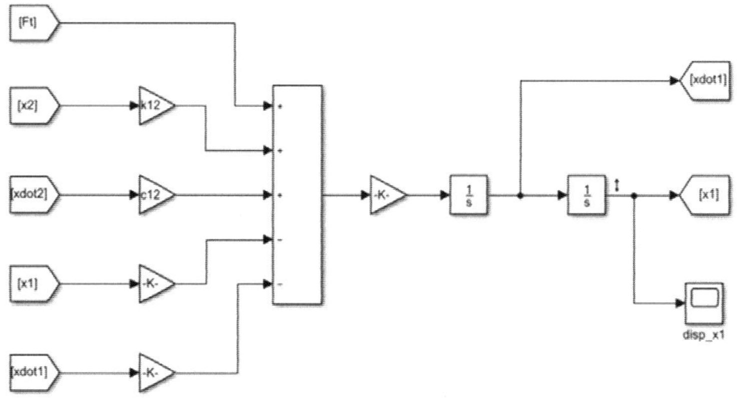

Figure 2. Simulink model for M1 (Head)

3. Results and discussion

Frequency Response Estimation and the bode plot for the simulation was obtained from the simulation. Due to large amount of data, only the Bode Plot for the head and neck are shown in Figure 3 and Figure 4. The results of the simulation are summarized in Table 3.

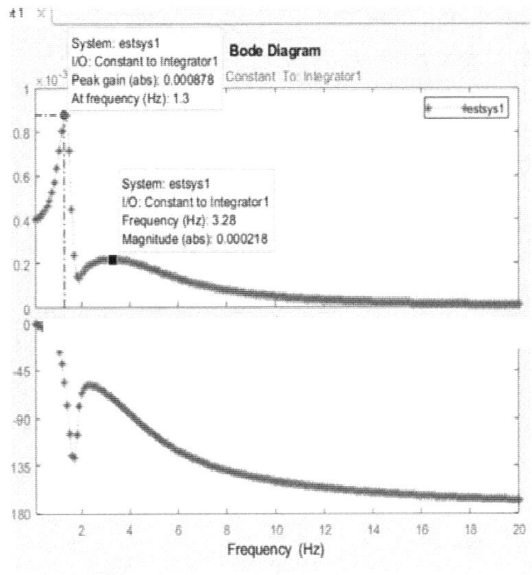

Figure 3. Bode Plot for M1 (Head)

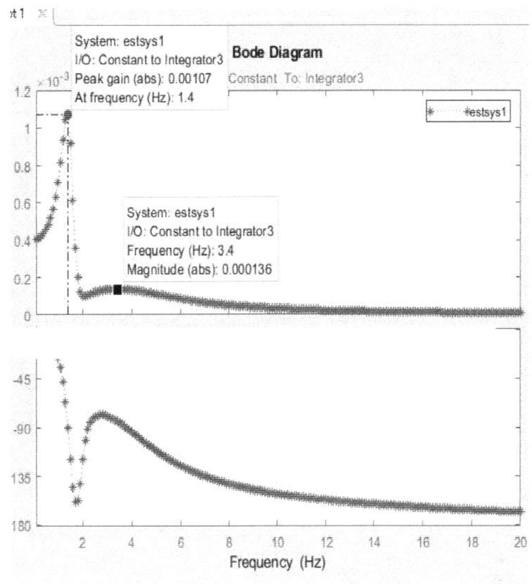

Figure 4. Bode Plot for M2 (Neck)

Table 3. Simulation results from Bode Plot

Mass	Description	Resonant Frequency (Hz)	Displacement Amplitude (m)
M1	Head [(1)]	1.30	0.000878
	Head [(2)]	3.28	0.000218
M2	Neck [(1)]	1.40	0.001070
	Neck [(2)]	3.40	0.000136
M3	Thoracic Spine	1.40	0.001280
M4	Lumbar Spine	1.40	0.001310
M5	Torso	1.40	0.001620
M6	Thorax	1.40	0.001620
M7	Diaphragm	1.40	0.001340
M8	Abdomen	1.40	0.001110
M9	Pelvis	1.40	0.001510

[(1)] Dominant Peak
[(2)] Secondary Peak

After the simulation, two peaks appeared in the Bode Plot for both head and neck. The first peak for the head (1.3 Hz) was rather significant and much more distinguishable. As for the second peak (3.28 Hz), it was rather diminished but still distinguishable. Similarly, for the neck, two resonant frequencies, 1.4 Hz and 3.4 Hz were found. The case was similar to the head where the first peak was more distinguishable than the second.

The second resonant frequency of the head (3.28 Hz) had a similar finding to [3], where the experiment conducted in his work showed the resonant frequency of the head to be 3 Hz. The slight difference may be due to the different constraints applied to the neck. In his experiment, the head was supported by a collar which further constrains the movement of the head. In this simulation, the head was assumed to be resting on a pillow. However, the results were logical as the values were quite close to the experiment.

The relation between the head and the neck is very close as they give almost similar responses from the simulation, showing a significant first peak and a distinguished second. The resonant frequencies of the head and the neck were also quite close for both peaks. A possible reason for this is that the head is directly connected to the neck in the body, which explains the similar responses. Another finding is that the second peak was only found in the head and the neck, but not the other body parts.

The results from the simulation would not be suitable to compare to the studies of vertical standing or sitting vibrations, as the position of the head, and other parts of the body, differs from the recumbent posture. This is because the weight of the body parts in the recumbent position are less likely to act on the connected parts, unlike in the vertical position, where the weight of the head rests on the neck and so on. Hence, the resonant frequencies found in this simulation differs greatly from vertical WBV.

3.1. *Resonant Frequencies for M3 to M9*

The resonant frequencies were similar for these masses, which was 1.4 Hz. This value is reasonably close to the findings of [5], which was 1.54 Hz for lower arms where the vibrational input was from the seat and injected horizontally to a seated human mathematical model. It was mentioned that when the vibrational input was from the seat, the vibrations will have to travel through the pelvis, abdomen, torso and thorax before reaching the lower arms. Although the lower arms were not investigated in the project's simulation, the similar resonant frequencies obtain show that the body parts may act as a single mass when subjected to recumbent vertical vibrations.

The value of 1.4 Hz was also found in the first peak of the neck (cervical spine). This similarity may be due to the rigid connection of the body parts (spine, torso, abdomen) when in recumbent state. The rigid connection may affect the response of these body parts to act as one single mass. This can be further observed from the displacement amplitude during resonance, where the displacement values differ slightly from one to another, especially in the case for torso and thorax.

The value of 1.4 Hz could be taken as the overall natural frequency of the body in the recumbent state as it corelates to 8 out of 9 parts of the body (excluding the head).

The results from the simulation are difficult to compare to previous studies of the vertical standing and sitting vibration as these are two completely different scenarios. However, the values of these resonant frequencies are logical as they do not differ much from the head and neck.

3.2. *Amplitude Responses*

Based on the 1.4 Hz peak, the lowest amplitude reading was taken from the neck (0.00107 m) while the highest amplitude was taken from the torso and thorax (0.00162 m), which differs only by 0.00055 m.

However, the neck allows more rotational movement compared to the other parts of the body. If the neck was excluded, the lowest amplitude reading would be taken from the abdomen (0.00111 m), which is considered to be a "softer" body part. Meanwhile, the head had the lowest displacement amplitude in the 1.3 Hz peak which was 0.000878 m. A possible explanation for this low amplitude reading may be the rotational effects of the neck and head.

Another possible reason for the similarity for the 1.4 Hz found in the neck was because the neck was part of the spine, which explains the similar vibrational characteristics to the thoracic and lumbar spine. The amplitude readings for the thoracic and lumbar spine are 0.00128 and 0.00131 m respectively, which only the differed by 0.00003 m. The slightly greater difference found in the neck can once again be explained by the fact that it provides more rotational movement compared to other parts of the body.

For the second peak of the head and neck (3.28 and 3.4 Hz), the displacement amplitude diminished to 0.000218 and 0.000136 m respectively. The amplitude of the head was much greater than the neck (deviation of 0.000082 m), which showed an increase of 60% from the neck to the head. This can be explained by the fact that the head is allowed to rotate while pivoted to the neck. Hence, the vertical displacement of the head is greatly enhanced as it is not only affected by the lateral stiffness of the neck but also the free rotational movement provided from the neck. This scenario could be addressed in future studies by including rotational stiffness in the simulation as done by previous studies. [8]

4. Conclusion

A mathematical model for a recumbent WBV simulation was developed for this research. The model was adapted from [5] and was modified accordingly for recumbent WBV. Also, a Simulink model was constructed from the model for the project's simulation. Resonant frequencies of the major body parts were obtained from the simulation and was recorded in Section 3. The data was captured in the form of Bode Plot where the resonant frequencies and amplitude responses were extracted.

The results from the simulation were more or less expected as they did not deviate greatly from previous works. Comparisons were made mainly based on the study of [3], as his research was on recumbent WBV. Comparisons were also made based on the findings of [5], where his simulation was on a seated model subjected to horizontal vibrations. The data obtained from the simulation differed slightly to previous studies due to unsimilar conditions. In the study of [3], the movement of the head was constrained by a neck collar. In the study of [5], the study was mainly to obtain the response of the lower arm. However, the resonant frequencies did not deviate much from the project's findings.

5. References

[1] Lundström, R. *et al.* (2018) 'Long-term effect of hand-arm vibration on thermotactile perception thresholds', *Journal of Occupational Medicine and Toxicology*. Journal of Occupational Medicine and Toxicology, 13(1), pp. 1–8.

[2] Duarte, M. L. M. H. and Matheus, D. B. P. (2006) 'Vision influence on whole-body human vibration comfort levels', *Shock and Vibration*, 13(4–5), pp. 367–377.

[3] Meusch, J. C. (2012) 'Supine human response and vibration-suppression during whole-body vibration', p. 78.

[4] Kimura, H. *et al.* (2017) 'Mechanical Bed for Investigating Sleep-Inducing Vibration', *Journal of Healthcare Engineering*, 2017, pp. 1–9.

[5] Qassem, W., Othman, M. O. and Abdul-Majeed, S. (1994) 'The effects of vertical and horizontal vibrations on the human body', *Medical Engineering and Physics*, 16(2), pp. 151–161.

[6] Yang, J. (James), Xu, P. and Kumbhar, P. (2008) 'Evaluation of Damping Coefficients

of a Vibratory Model of Human Body', *Mechanical Engineering.*

[7] Panjabi, M. M. (1973) 'Three-dimensional mathematical model of the human spine structure', *Journal of Biomechanics*, 6(6), pp. 671–680.

[8] Knez, L., Slavič, J. and Boltežar, M. (2017) 'A Sequential Approach to the Biodynamic Modeling of a Human Finger', *Shock and Vibration*, 2017, pp. 1–12.

Acknowledgement

The authors would like to express their gratitude to all who have participated in this research. The support from the Department of Mechanical Engineering, Faculty of Engineering and Technology (FOET), Tunku Abdul Rahman University College (TARUC) is fully acknowledged.

IOP Conf. Series: Materials Science and Engineering **815** (2020) 012016 doi:10.1088/1757-899X/815/1/012016

Investigation of Fatigue Failure Performance of Ankle Implant Mobile Bearing using Finite Element Method

Rosdi Daud[1*], M Z H Jantan[1], H M Ayu[1] and A Shah[2]

[1]Faculty of Mechanical and Automotive Engineering Technology, Universiti Malaysia Pahang, 26600 Pekan, Pahang, Malaysia
[2]Faculty of Technical and Vocational, Universiti Pendidikan Sultan Idris, 35900 Tanjung Malim, Perak, Malaysia

*Corresponding author: rosdidaud@ump.edu.my

Abstract. The Fatigue of the mobile bearing component of ankle implant became one of the main causes of failure in ankle implant. This paper deal with the investigation of fatigue failure performance of mobile bearing for different gait cycles (normal, dorsiflexion and plantarflexion) using finite element method. The finite element analysis is one of the method to predict fatigue and identify the critical area that have highest contact stress on mobile bearing design of ankle implant. The three-dimensional solid modelling of STAR and BOX ankle implants are constructed using SolidWorks software. The finite element analyses are performed by using ANSYS software. The finite element model of the mobile bearing was analysed using static structural analysis approach. The most critical area on the mobile bearing of ankle implant was found on its middle bottom area. The obtained results also showed that the STAR mobile bearing has the higher life cycle than BOX before fatigue failure occurs. This finding is similar with experimental and clinical result done by previous researcher. Therefore, our Finite Element Model has potential to improve the mobile bearing at the designing stage to be better in future in term of fatigue failure resistance and has longer life span.

1. Introduction

Nowadays, fatigue of the mobile bearing component of ankle implant became one of the main causes of failure in ankle implant which is also lead to its loosening. Loosening of the mobile bearing component of ankle implant also occur because of wear. Previous research did study on the prediction of wear on the mobile bearing component [1, 2]. Wannomae et.al reported that the cyclic contact stress at articular surfaces such as between the polyethylene mobile bearing and metal components in ankle implant, mobile bearing undergoes pitting, delamination and changes in crystal structure resulting in low resistance to wear [3]. The current designs of the ankle implant needs a new improvement or innovation to improve the lifespan of the devices [4]. There has been a renewed interest in several new design of ankle implant component after several disappointing in clinical results of the earlier ankle implant designs [4, 5]. The first type of polyethylene used as mobile bearing material was ultra-high molecular weight polyethylene (UHMWPE) which was popularised by Sir John Charnley in the 1960s with his Low Friction Arthroplasty (LFA). Coupled with a metal bearing, UHMWPE provides a high rate of satisfaction, outcome and survivorship in both total hip replacement and total knee replacement. However, aseptic loosening is dominating ankle implant failures and revision [6]. In reducing the risk of aseptic loosening, UHMWPE is used as a liner material in ankle implant since 1960s. It is maybe due of its superior mechanical properties like high strength, low creep, low friction coefficient and good resistance to fatigue [7]. The wears of polyethylene leading to osteolysis in long term period due to the development of wear particles which cause bone losses surrounding implants leads to instability and subsequently loosen of the implant components [8]. Major factors that contribute to failure of ankle implant such as are fixation method and component design [9]. In order for the ankle implant to have better function and improvement, it must be good in fundamental to enhance the understanding of postoperative performance. The knowledge is mainly on clinical and radiological assessments and survival rates [10]. However, knowledge to enhance understanding the ankle implant design also can be obtained by investigation using finite element method as be done by many previous researchers in investigation other implant performance or biomechanics [11-15]. Nevertheless, the approach of using finite element method to simulate the life cycle for different gait cycle in mobile bearing components is still lacking. Thus, this paper aim is to investigate the fatigue failure performance of mobile bearing for different gait cycles (normal, dorsiflexion and plantarflexion) using finite element method.

2. Methodology

The metals Co-Cr-Mo (ASTM F-75) was used as tibial and talar components for all the ankle implant models and highly cross-linked UHMWPE was used for the mobile bearing component. As we know the Co-Cr-Mo alloys have great resistant to pitting and crevice corrosion and these alloys have excellent wear resistance properties [16]. Table 1 shows the mechanical properties of ankle implant components.

Table 1. Mechanical properties of ankle implant components.

Properties	Mobile bearing (UHMWPE)	Tibial/Talar (Co-Cr-Mo)
Density (kg/m^3)	940	8768
Young Modulus (MPa)	557	210000
Tensile Strength (MPa)	40	655
Yield Strength (MPa)	25	450
Poisson Ratio, V	0.45	0.29

Finite element analysis recently used to assess the influence on contact pressure and internal stress distribution. An average male body weight of 732 N considered for this study with simple loading conditions applying in static loading. The load was remained the same as the force exerted on the surface of the tibial to generate stress distribution on the mobile bearing. The ankle implant was simulated using different gait cycle by applying dorsiflexion (10° angle) and plantarflexion (25° angle). In case of fixed bearing model, the mobile bearing fixed to the tibial component in neutral position by using 'connector' tool in ANSYS software.

2.1. Geometric Model

A three-dimensional (3D) model of an ankle implant was built using the SolidWorks software. By using SolidWorks software, the 3D model of STAR and BOX implant were constructed based on the dimensions obtained from the previous journal. The STAR and BOX implant were chosen because of the three components were congruence with physiologic ankle mobility. Figure 1 shows the 3D model of talar component, mobile bearing and tibial component of STAR implant.

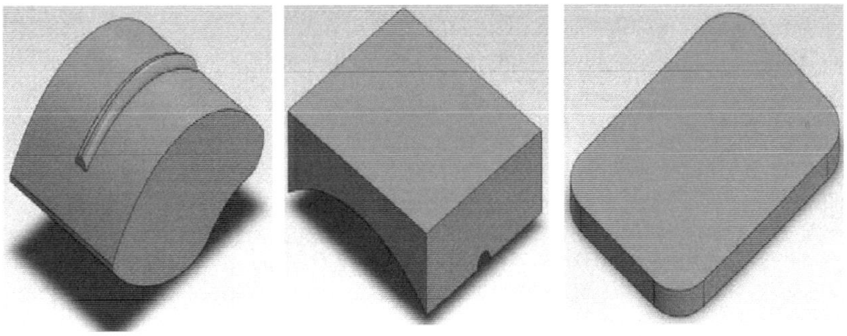

(a) (b) (c)

Figure 1. (a) STAR Talar implant (b) STAR mobile bearing and (c) STAR tibia implant

After all three-components of STAR implant was constructed, then it was assembled together by using SolidWorks software. Figure 2 shows completed 3D model of STAR implant.

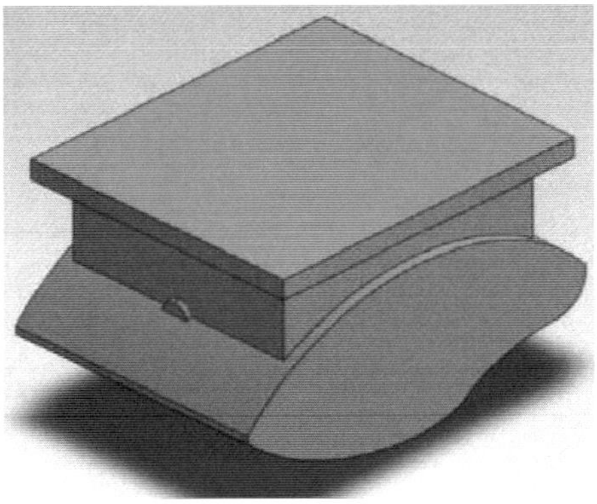

Figure 2. 3D Model of completed STAR ankle implant

Figure 3 shows the 3D model of talar component, mobile bearing and the tibial component of BOX implant.

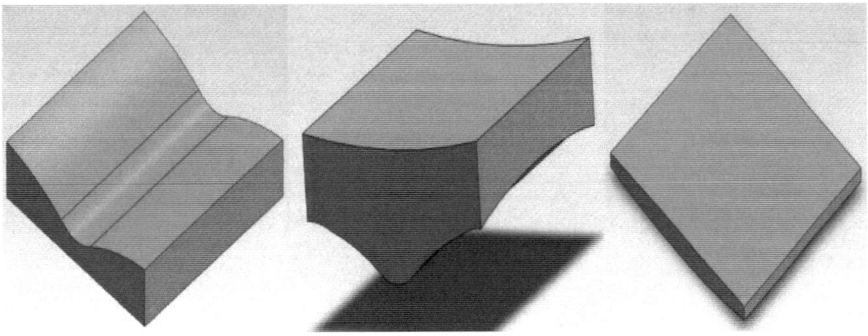

(a) (b) (c)
Figure 3. (a) BOX Talar implant (b) BOX mobile bearing and (c) BOX tibia implant

All these models were developed to simulate the right ankle implant. The metal part of tibial and talar components were assigned as Cobalt-Chromium alloy (Co-Cr-Mo) material properties where the Young's modulus of 210 GPa and Poisson ratio of 0.29 were applied to the model. The mobile bearing component of STAR implants have a concave of 25 mm radius and both bearing on the inferior surface are intact with talar components [17]. The bearing component is ultra-high molecular weight polyethylene manufacture from GUR 1020 with Young's modulus of 557 MPa and Poisson ratio of 0.49.

2.2. Finite Element Analysis of Gait Cycle

The finite element analysis is very useful in predicting the fatigue failure on the mobile bearing. The finite element analysis was done using ANSYS software to get the contact stress in determining the maximum life cycle of the mobile bearing during different gait cycle (normal, plantarflexion and dorsiflexion). The finite element analysis was used to visualize the fatigue failure behavior of the mobile bearing for both STAR and BOX implant.

2.3. Apply Loads and Boundary Conditions

The tibial component was applied the force from the bone and the talar component was fixed at the articulating surfaces of the talar bone. In this simulation, the load created was an axial (vertical) load using concentrated force applied to the tibial component. Figure 4 show the axial force applied to the tibial component of ankle implant.

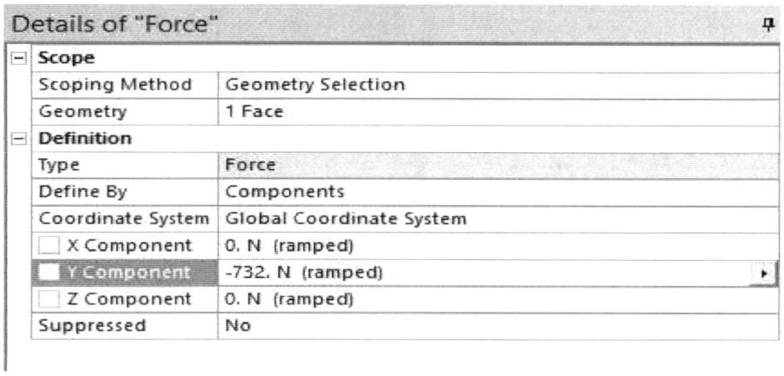

Figure 4. Force applied to the tibial component of ankle implant.

As can be seen from figure 4, the value of load -732 N was applied as the load direction going downward in y-axis while figure 5 shows the fixed part at the bottom of the talar component.

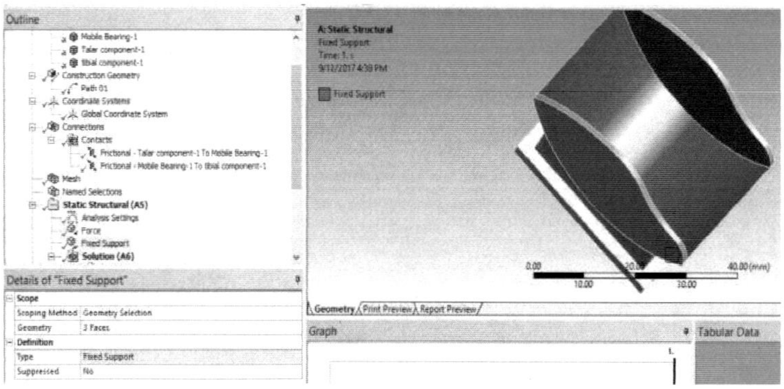

Figure 5. The fixed at the bottom of the talar component

In finite element simulation, the concept of contact and target surface is used for each contact region. One side of a contact region is referred as the contact surface while the other side is referred as target surface. The contact surfaces are restricted from penetrating through the target surface. When one side is designated the contact and the other side is target this is called asymmetric contact. Figure 6 shows contact surface of mobile bearing with tibial component.

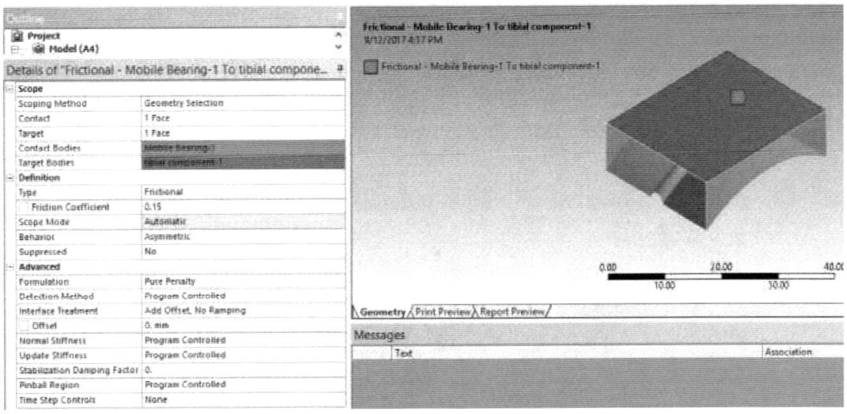

Figure 6. shows contact surface of mobile bearing with tibial component.

From figure 6 show the assemblies in frictional body contact with 0.15 frictional coefficient for both contact of mobile bearing, talar and tibial.

2.4. Mesh the Model

There are hexa, wedge and tetra elements for the solid mesh element available in ANSYS. Tetrahedral element is used in automatic meshing algorithms. Figure 7 shows the mesh of mobile bearing using ANSYS software.

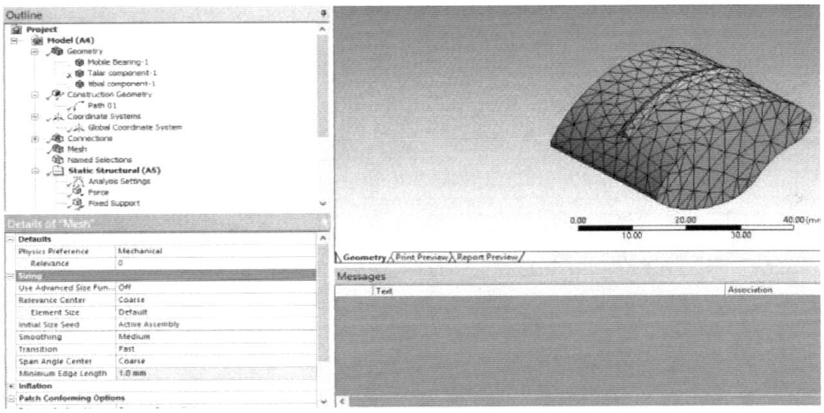

Figure 7. The mesh of mobile bearing using ANSYS software

3. Result and Discussions

Based on FE simulation results obtained as shown in figure 8, the stress Von Mises shows maximum stress of 20.939 MPa exerted at the bottom surface of the mobile bearing during normal gait cycle. During normal gait cycle most of the force exerted on the middle area of the mobile bearing. While, the stress Von Mises shows maximum value 17.603 MPa exerted at the bottom surface of the STAR mobile bearing during dorsiflexion. Most of the force exerted on the upper area of the mobile bearing.

IOP Conf. Series: Materials Science and Engineering **815** (2020) 012016 doi:10.1088/1757-899X/815/1/012016

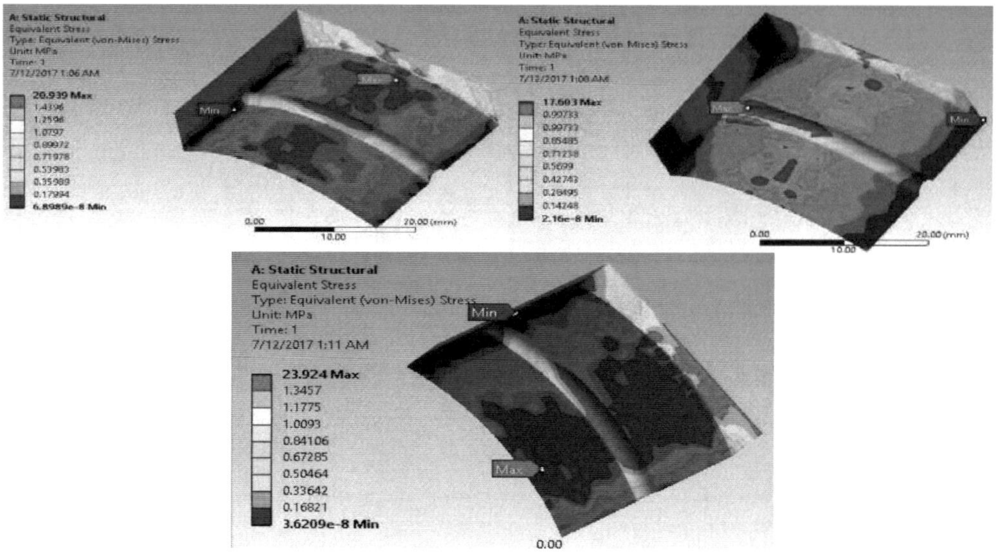

Figure 8. The FE simulation results (stress von-mises) for STAR implant during normal, dorsi flexion and plantarflexion gait cycle

The stress values reach the peak, which is 23.924 MPa during plantarflexion. Most of the force exerted on the lower area of the mobile bearing. However, the stress value during normal, dorsiflexion and plantarflexion are still under the yield strength, thus ensured that the stress relaxation of the material is within acceptable limit. Based on FE results obtain on BOX implant, the figure 9 show the stress Von Mises of 53.962 MPa exerted at the bottom surface of the mobile bearing during normal gait cycle. While, during dorsiflexion gait cycle, stress Von Mises maximum value of 31.129 MPa exerted at the bottom surface of the BOX mobile bearing. The highest value of stress Von Mises, 40.439 Mpa can be seen exerted at the bottom surface of the BOX mobile bearing during plantarflexion. By the way, most of the force exerted on the middle area of the mobile bearing which is can be claimed the most critical region on the mobile bearing.

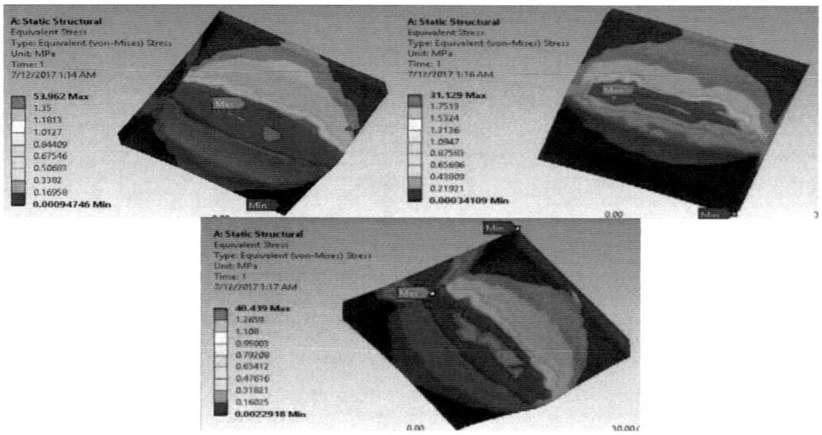

Figure 9. The FE simulation results (stress Von Mises) for BOX implant

IOP Conf. Series: Materials Science and Engineering **815** (2020) 012016 doi:10.1088/1757-899X/815/1/012016

Based on the life cycle result obtained as shown in figure 10, the minimum life cycle of 1213.9 is at the bottom surface of the mobile bearing. As we can see, during normal gait, most of critical area is occur on the middle area of the mobile bearing. While, the life cycle result shows minimum life cycle of 941.24 at the bottom surface of the mobile bearing. As we can see, during dorsiflexion, most of the critical area is occur at the middle to upper area of the mobile bearing. The shortest life cycle, 776.87 can be seen at the bottom middle area of the mobile bearing during plantarflexion.

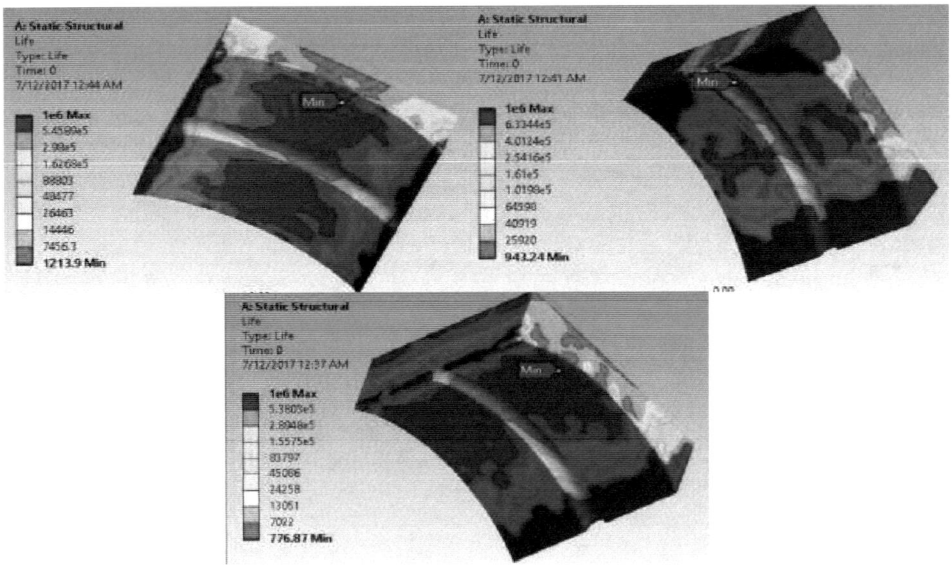

Figure 10. The FE simulation results (fatigue life cycle) for STAR implant

The S-N curve (fatigue limit) of the STAR mobile bearing during plantarflexion is as shown in figure 11. Fatigue limit for plantarflexion is chosen instead of normal or dorsiflexion gait cycle is due to the highest stress value is found during plantarflexion gait cycle.

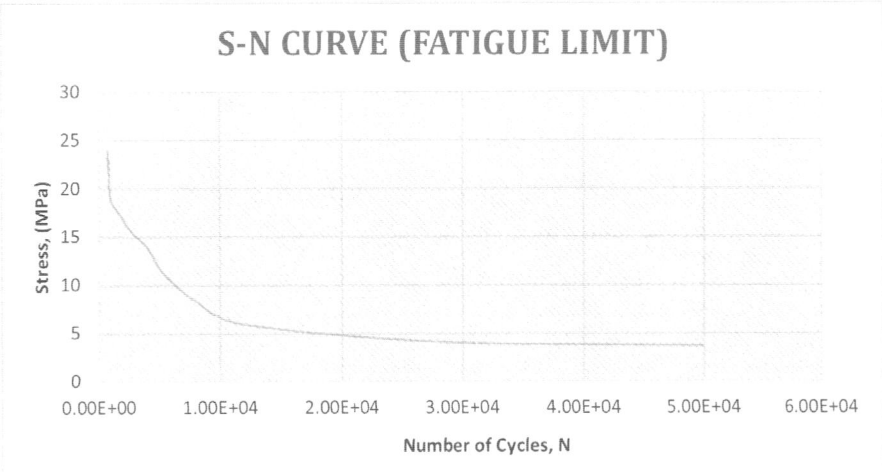

Figure 11. The FE simulation results (fatigue life cycle) for STAR implant

Based on the data shows in figure 11, during peak stress at 23.924 MPa the lowest cycle is 776.87. This finding is in agreement with the previous research that reported due to relatively low yield point and high wear resistance of UHMWPE, the contact pressure acting on the surface causes higher fatigue failure rates in ankle implant [18]. When the higher loads acting on the ankle joint during gait cycle, it generate higher contact pressure which leads to ankle implant failure. Overall, refer to Table 2 we can see that the lowest life cycle of mobile bearing is clearly happen during plantarflexion gait cycle. While, the STAR implant gave highest life cycle of mobile bearing which almost 80% longer than BOX implant. This is maybe due to STAR mobile bearing design is in concave shape, which makes it larger surface area than BOX mobile bearing. While, based on mating surface principle the larger surface area will distribute the loads at the contact to a large amount of body material and thereby induce a more uniform stress pattern.

Table 2. Comparison of Life Cycle of Mobile Bearing between STAR and BOX implant

Life cycle of Mobile Bearing	STAR	BOX
Normal	1213.9	331.78
Dorsiflexion	943.24	243.77
Plantarflexion	776.87	172.67

The success of an ankle implant depends on the design of the implant's mobile bearing itself to overcome fatigue failure to occur. Fatigue is one of the main failure that caused the failure of the mobile bearing on the ankle implants due to cyclic motion in stance phase. Direct comparison of the present results with previous research results is difficult because there is lack of analysis done on fatigue performance of ankle implant mobile bearing.

4. Conclusion

The main objective of this study was achieve since our finite element model able to predict the most critical area which is fragile to the cyclic load during gait cycle (normal, dorsiflexion and plantarflexion). Our FE model is also able to predict life cycle of mobile bearing for both most popular ankle implant, STAR and BOX. Thus, based on the finite element simulation result obtained, we can concluded that the BOX implants have higher potential for the fatigue failure to occur than the STAR implant due to its highest stress approximately 53.96 MPa although during normal gait cycle. If the mobile bearing fails, it leads to the loosening of the whole ankle implant. While, STAR implant showed that it is a reliable choice to use since it has higher life span.

5. References

[1] A.M.S. Putra et al. 2014 *Advanced Materials Research* **845**
[2] Nicolo Martinelli, Silvia Baretta, Jenny Pagano, Alberto Bianchi, Tomaso Villa, Gloria Casaroli & Fabio Galbusera 2017 *BMC Musculoskeletal Disorders* **18**
[3] Wannomae, K.K., Christensen, S.D., Freiberg, A.A., Bhattacharyya, S., Harris, W.H., Muratoglu, O.K. 2006 *Biomaeterials* **27**.
[4] Kathryn O'Connor, Sandra Klein, Patrick Ebeling, Adolph Samuel Flemister, Phinit Phisitkul 2018 *Foot and Ankle Ortho-paedics* **10**
[5] Cracchiolo A, DeOrio JK 2008 *J Am Acad Orthop Surg* **16**
[6] S. Giannini, M. Romagnoli, J. J. O'Connor, F. Catani, L. Nogarin, B. Magnan, F. Malerba, Massari, M. Guelfi, 2011 *Journal of Foot and Ankle Surgery* **50** pp 641-647
[7] Li, S., Burstein, A. H, 1994 *J Bone Joint Surg Am* **76** 1080
[8] E. Ingham and J. Fisher 2005 *Biomaterials* **26** pp. 1271-1286.
[9] Nishikawa, M.,Tomita, T., Fujii, M. Watanabe, T., Hashimoto, J., Sugamoto, K., Yoshikawa, H. 2004 *Int. Orthop.* **28** pp. 123-126.

[10] Stengel D, Bauwens K, Ekkernkamp A, Cramer J 2005 *Arch Orthop Trauma Surg* **125**, pp. 109–119
[11] EL Sin Jian, Rosdi Daud, HM Ayu, MH bin Yusoff, J Jamiluddin, A Shah 2019 *IOP Conference Series: Materials Science and Engineering* **469**
[12] NMA Azam, Rosdi Daud, H Mas Ayu, J Ramli, MFB Hassan, A Shah, MAHM Adib 2018 *MATEC Web of Conferences* **225**
[13] Rosdi Daud, Sulaiman Suaidah, H Mas-Ayu, Siti Haryani Tomadi, MS Salwani, Arman Shah, Mohammed Rafiq Abdul-Kadir 2018 *Materials Science Forum* **916** pp.153-160
[14] FA Zakaria, R Daud, H Mas Ayu, SH Tomadi, MS Salwani, Mohammed Rafiq Abdul Ka-dir 2017 *Matec Web of Conferences* **108**
[15] Li-Xin Guo, Jia-Yu Yin 2019 *Medical & Biological Engineering & Computing* **57** pp 89–98
[16] Mitsuo Niinomi 2010 *Metals for Biomedical Devices* (Woodhead Publishing)
[17] Leardini, J. J. O'Connor, F. Catani and S. Giannini *2004 Clin. Orthop* **424**
[18] Reggiani B, Leardini A, Corazza F, Taylor M. 2006 *Journal of Biomechanics* **39**

Acknowledgement
We would like to thank Universiti Malaysia Pahang through research grant RDU180395 for fully support the facilities and resources for this research.

Effect of Femur Geometry on Hip Protectors Testing System

S A Yahaya[1,2]*, A A S Ahmad Iqbal[1], Z M Ripin[1] and M I Z Ridzwan[1]

[1]School of Mechanical Engineering, Universiti Sains Malaysia, Penang Malaysia
[2]Department of Biomedical Engineering, University of Ilorin, Ilorin, Nigeria.

*Corresponding author: yahaya.say@student.usm.my

Abstract. There have been different types of simplified femoral geometry incorporated into test systems for testing the biomechanical effectiveness of hip protectors. However, the effect of the simplification of femoral geometry in simulated sideways impact experiment has not been reported. In this study, an actual femoral geometry was fabricated and examined along with a simplified femoral geometry in a surrogate hip arrangement to mimic the fracture-causing situation in a sideways fall. A pendulum impact testing machine was used to evaluate the impact force response of both the femoral geometries at various velocity and constant residual impact energy for both geometries. The result showed that the peak impact force varied only 9% between actual and simplified geometries after normalizing the weight for both femurs in testing without a hip protector. However, a higher distinction of the impact force responses of both femurs was observed when testing with a hip protector. This findings at this moment suggest that the femur geometry affects the performance of a testing system. Therefore, the simplified femur geometry may not accurately substitute for the actual femur geometry in comparison of the impact attenuation of a hip protector in a simulated sideways fall.

1. Introduction

The response of a surrogate hip assembly is a critical factor in hip protector testing. The femur geometry represents a key component in the hip assembly. Its weight (effective mass) has been proven to affect the output force in a simulated impact experiment [1]. Though, this may include the anatomical features of such geometry. The surrogate femur, which is responsible for the dynamics of impact force response in vitro representation of real fall, has a considerable role to play in hip protector testing systems [2].

Drop tests are usually carried out to analyze energy dissipation of hip protectors in dynamic load setting with employed femoral geometry having the ability to produce loads that could be concentrated or distributed depending on the simulated model of the impact surface and/or simulated nature of the hip bone [3]. The need to simplify human anatomy in a test rig because of the difficulty in modelling irregular and complex geometry of biological system like the femur must not be used as a factor to compromise essential fracture causing events. Therefore, employed geometry must satisfy the condition to offer the expected response of a hip impact in a sideways fall.

Nevertheless, it has been argued that many parts of the anatomy of the hip structure, as demonstrated by the deformation of the soft organs and the articulation of the femur and pelvis, may not contribute to the magnitude of the impact force of the trochanter [4], the nature of impact site may have significant effect as demonstrated by disparities in reported values of hip protector performances using different test systems [2].

Building on the representation of trochanter with a semi-circled striker by Bulat et al. [5] and the axially symmetrical ellipsoid on a pendulum contact surface used to represent the greater trochanter by Parkkari et al. [1]. Haris et al. [3] had used a flat and a ball impactor to demonstrate the effect of the roundness of the hip bone under impact configuration and showed distinction in the forces transmitted against a case when the roundness was jettisoned, though Haris et al. [3]had slight differences in the energies used in the evaluation of both configurations, the bare impact forces were representative of earlier reported values by Wiener et al. [6]. Additionally, in a bid to represent repeatable measuring method in a laboratory-controlled environment, Keenan and Evans [7] used anatomic femur model that could also pass for a similar geometry of a semicircular feature with a given radius to represent the greater trochanter but also having a distal part as though portion of the femoral shaft had been modelled. This anatomic model could be seen as a build upon the rod used by Mills [4] to represent the femur.

While there exist various impact test systems that do not incorporate the anatomical shape of the femur; using mechanical anvil or a hip shaped anvil for testing as the representation of the pelvis [8], a number of work incorporate the actual geometry of the femur in their surrogate hip model [9–12]. These differences in femur geometries may affect the hip protector testing system [2]. Towards, the standardization of the test systems, it may be interesting to evaluate how a simplified femur geometry response differs from that of the actual geometry.

In this study, we evaluated the impact response of the actual femur model against a simplified model, in vitro, in simulated impact to the sideways.

2. Methodology Formatting the title, authors and affiliations

2.1. Experimental Setup

Two types of femur geometry were fabricated; simplified and actual bone geometry (Figure 1). The simplified model of the femur bears the profile of the greater trochanter as traced from a Sawbones 4th generation computer-aided design (CAD) model in SolidWorks 2016. The Aluminium block (195 x 40 x 50) mm was machined by using a 3-axis computer numerical control (CNC) milling machine (Robodrill α-T21iFLb). The simplified femur geometry was based on the geometry employed by Keenan and Evans [7]. Meanwhile, the actual geometry of femur bone was fabricated using a 5-axis CNC milling machine (DMU 40 Monoblock). Aluminium was used to represent the bones because it has lightweight, it can effectively transmit forces and it is resistant to high impact force usually employed in an impact test.

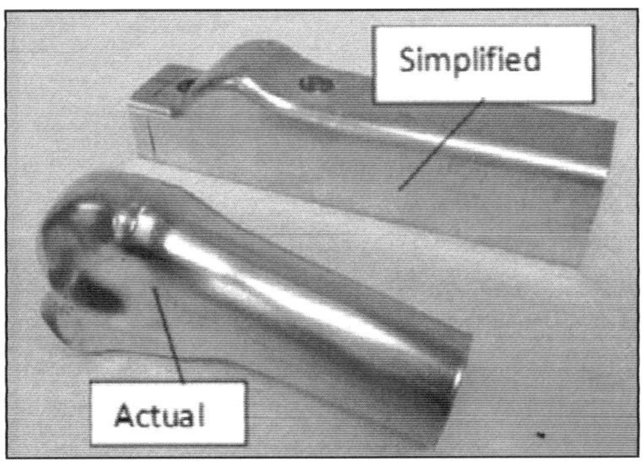

Figure 1. The actual and simplified femur geometry

The femurs were incorporated into a surrogate pelvis that was attached to an impact pendulum (Figure 2) which allowed impact from various inclined positions to hit a load cell (20kN capacity Kistler Model CH-8408) that was connected to a Multichannel Charge Amplifier (Type 5070) and IMC device data logger. The surrogate pelvis comprised of simulated human soft tissue made from polyethylene foam [14] and proximal femur as earlier described. A pendulum impact machine 70 cm in length was fabricated for the experiment. It has a protractor that is attached to the pendulum holder to indicate the impact angle and incorporated adjustable slots for mass up to 30 kg. The study also investigated the biomechanical testing on soft hip protector (HipSaver) with a diameter of 19 cm and 1.8 cm thick. It is made of foam material and layered with a thin layer of plastic.

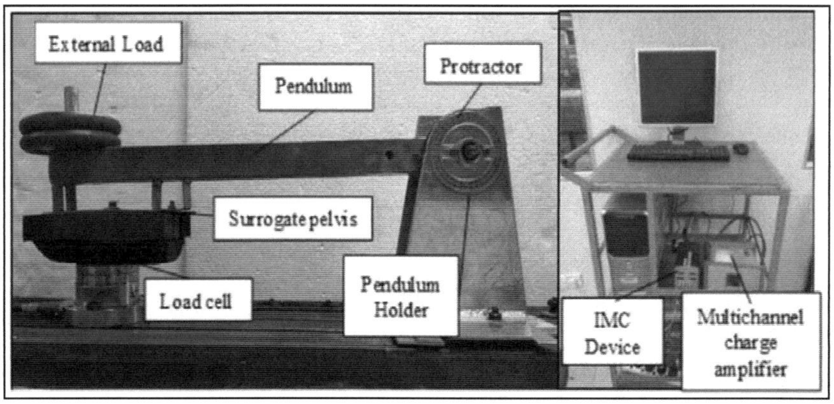

Figure 2. (a) The pendulum hip impact testing system (b) The Multichannel charge amplifier, IMC device and desktop used to record the data received from the Kistler load cell

2.2. The impact tests

The impact testing for this experiment was carried out under the following conditions:

- No hip protector on simplified and actual femur geometries without external load.
- No hip protector on simplified and actual femur geometries with 2.5kg load.
- No hip protector on simplified and actual femur geometries with 5kg load.
- Soft hip protector condition for the above-mentioned cases (1 -3)

Trials were performed at impact angles of 10°, 20°, 30°, 40°, 50°, 60°, 70°, and 80°. Twenty-four trials were carried out for each type of femur bones. T-test was further used to observe whether the impact force in each set of two test conditions differed significantly at a p-value of 0.05.

3. Results and discussion

Each figure should have a brief caption describing it and, if necessary, a key to interpret the various lines and symbols on the figure.

3.1. The effects of simplified and actual femur geometries on impact force

The result of the effect of femur geometry in terms of normalized impact force response is presented in Figure 3. As shown in the figure, there are two significant behaviours noticed; firstly between 1.13 – 2.53 kN simulated impact on actual geometry, the actual femur had a higher impact force response of 0.91 kN (±1.08) and secondly between 2.56 – 10.67 kN, the simplified femur had a higher impact response of 1.29 kN (± 0.78). The second range is particularly of interest as most hip fracture studies examine impact above 2.5 kN and less than 10 kN. When the condition was adjusted to cause an increase in the impact force response of the actual geometry, an increase in the impact force response of the simplified geometry was also observed (r2= 0.962), but the relationship is not linear. A nonlinear relationship exists between the two femurs, which suggest an unpredictable situation in using one to represent the other. All the impact forces show different response pattern though the peak forces increase with impact angle and load. As expected, the total impact time for both cases decrease as the height of impact increases. The result of t-test reveals that the impact force response of the two geometries significantly differs (p-value < 0.005).

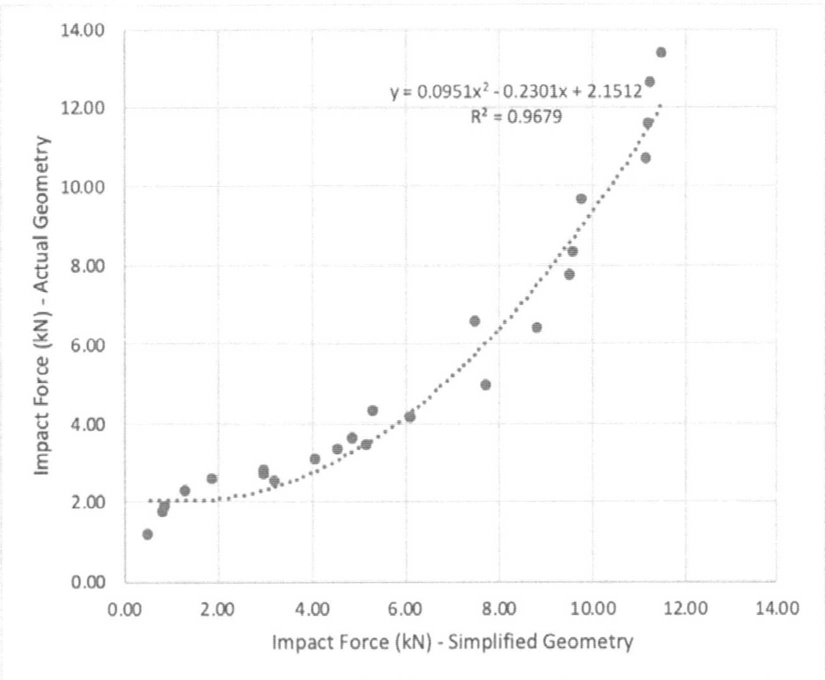

Figure 3. The difference in impact force between the simplified and the actual geometry of the femur.

3.2. Evaluation of the impact attenuation of the hip protector

The difference in the response of the femurs incorporated into the surrogate hip became more apparent by using both arrangements to test the impact attenuation of the soft hip protector (Figure 4). As the impact forces increase from low to high, the impact attenuation of the hip protector when the actual femur was used for the testing maintained a relatively linear curve while the response became parabolic when the simplified femur geometry was employed. Although the impact happening on the same kind of surface, since the hip protector overlaying the hip shaped soft tissue is the surface that comes in contact with the impact plate, the underlying femoral geometry plays a significant role in the force received by the hip in a simulated sideways fall thereby contributing to the debate on the effect of the test system on the performance of hip protectors. It is at this moment, suggested that care must be taken in implementing the simplification of this complex system to avoid simulation that may be unrealistic in actual fall condition.

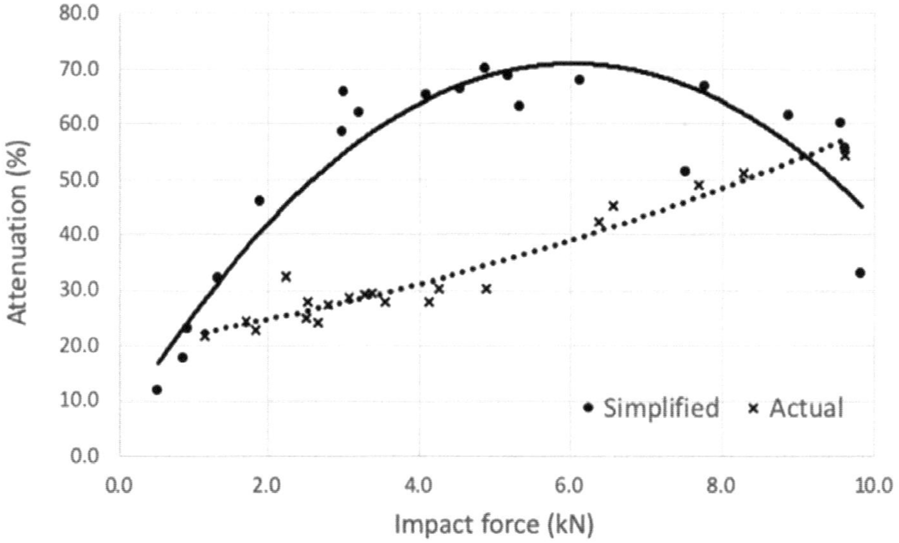

Figure 4. The impact attenuation provided by soft hip protectors with different types of femur geometry.

4. Conclusion

The impact force between the simplified and actual femur geometry, at the lower energy conditions, is slightly different, however as the conditions were varied to increase the impact force, the difference in impact forces between both femurs became apparent. After normalizing the weight for both femurs, there exist only 9 % difference in the impact force response between the actual and simplified geometry. The difference between the reaction of the geometries in hip protector testing showed a distinctive effect in the evaluation of the same hip protector under the same test condition. In conclusion, the simplified geometry may not be suitable to replace the actual geometry of a femur in an impact test system for the evaluation of the performance of a hip protector. Thus, models must closely mimic the real femur in impact force evaluation of a simulated sideways fall.

5. References

[1] Parkkari J, Kannus P, Poutala J and Vuori I 1994 Force attenuation properties of various trochanteric padding materials under typical falling conditions of the elderly *J. Bone Miner. Res.* **9** 1391–6

[2] Yahaya S A, Ripin Z M and Ridzwan M I Z 2019 Test systems for the biomechanical evaluation of hip protectors: a systematic review *Osteoporos. Int.* **31** 43–58

[3] Haris A, Goh B W Y, Tay T E, Lee H P, Rammohan A V. and Tan V B C 2018 On the effectiveness of incorporating shear thickening fluid with fumed silica particles in hip protectors *Smart Mater. Struct.* **27** 01–12

[4] Mills N J 1996 The biomechanics of hip protectors. *Proc. Inst. Mech. Eng. H.* **210** 259–66

[5] Bulat T, Applegarth S, Wilkinson S, Fitzgerald S G, Ahmed S and Quigley P 2008 Effect of multiple impacts on protective properties of external hip protectors *Clin. Interv. Aging* **3** 567–71

[6] Wiener S L, Andersson G B J, Nyhus L M and Czech J 2002 Force reduction by an external hip protector on the human hip after falls *Clin. Orthop. Relat. Res.* **398** 157–68

[7] Keenan B E and Evans S L 2019 Biomechanical testing of hip protectors following the Canadian Standards Association express document *Osteoporos. Int.* 1205–14

[8] Nabhani F and Bamford J 2002 Mechanical testing of hip protectors *J. Mater. Process. Technol.* **124** 311–8
[9] van Schoor N M, van der Veen A J, Schaap L A, Smit T H and Lips P 2006 Biomechanical comparison of hard and soft hip protectors, and the influence of soft tissue *Bone* **39** 401–7
[10] Schmid Daners M, Wullschleger L, Derler S and Schmitt K-U U 2008 Development of a new design of hip protectors using finite element analysis and mechanical tests *Med. Eng. Phys.* **30** 1186–92
[11] Laing A C, Feldman F, Jalili M, Tsai C M (Jimmy) and Robinovitch S N 2011 The effects of pad geometry and material properties on the biomechanical effectiveness of 26 commercially available hip protectors *J. Biomech.* **44** 2627–35
[12] Choi W J and Robinovitch S N 2018 Effect of pelvis impact angle on stresses at the femoral neck during falls *J. Biomech.* **74** 41–9

Acknowledgments
The authors would like to acknowledge technical assistance from Mr. Wan Mohd Amri Wan Mamat Ali, Mr. Faris Mohamed Zailan, Mr. Abdul Halim Che Aat, Mr. Baharom Awang, and Mr. Norijaz Abd. Aziz of the School of Mechanical Engineering, USM. The financial support provided under USM RUI grant (1001/PMEKANIK/8014070).

IOP Conf. Series: Materials Science and Engineering **815** (2020) 012018 doi:10.1088/1757-899X/815/1/012018

Modelling and Closed–loop Control System for Magnetorheological Elastomer Linear Actuator

M N Hazwan Hadzir[1], Muhamad Husaini Abu Bakar[1]*, Ishak Haji Abdul Azid[1]

[1]System Engineering and Energy Laboratory, Universiti Kuala Lumpur Malaysian Spanish Institute, Kulim Hi-Tech Park, 09000 Kulim, Kedah.

*Corresponding author: muhamadhusaini@unikl.edu.my

Abstract. Controlling the actuator in micro to microscale positioning is the challenge in the development of smart material actuator. The design of the existing commercial piezoelectric actuator is limited due to the material properties itself such as brittle which require special packaging and protection in order to avoid creep occurred. Magnetorheological linear actuator representing a relative new class in actuator development. the flexibility in MRE material potentially overcome the limitation of the piezoelectric actuator. This paper presenting the development of closed loop control system including the dynamic model, plant model and PID controller for linear MRE actuator. The control system was successfully developed and the MRE linear actuator able to generate the displacement at certain steady state error.

1. Introduction

The micro component product took place in the development of technology production nowadays. Due to the necessity, the high precision and accuracy of production keep emerging and become high demands. The actuators are closely related to the determination of accuracy and precision in the positioning system in order to achieve high precision and accuracy.

The actuator which uses smart material based was introduced to design the actuator in micro-scale motion [1] including the shape memory alloy, piezoelectric, and magnetorheological materials [2]. The shape memory alloy (SMA) is an alloy that able to restore its original shape after removed the heat [3]. The implementation of SMA as a damper to absorb vibration in the actuator application since their early development in the year 1988 but the operation of the SMA actuator is very complex due to the requirement of both heating and cooling system to produce the displacement.

Thus, piezoelectric material has become the most attractive smart material based for actuator [4] due to its simplicity. The displacement of the piezoelectric actuator deformed by external electric fields [5]. Piezoelectric actuators are promising to create nano-scale displacement [6]. However, the design of the piezoelectric actuator is limited due to the material properties itself such as brittle which require special packaging and protection in order to avoid creep occurred. Therefore, looking for another smart material to overcome the limitation of piezoelectric material is a need.

The magnetorheological (MR) material is also categorised in the smart material group. The displacement of the actuator was varied with magnetic field induced [7]. The common types of magnetorheological materials is magnetorheological fluid (MRF) and magnetorheological elastomer (MRE) [8]. The magnetorheological fluid is a type of intelligent fluid in a carrier fluid commonly in oil.

The properties of fluid will change to viscoelastic solid when external magnetic was applied. The universal industrial application-based magnetorheological fluid is damper, valve, clutch, and braking system [9].

Besides, the magnetorheological elastomer, which also known as a magnetosensitive elastomer, is the class of solids that consist of the polymer matrix with embedded micro or nano-sized ferromagnetic particle such as carbonyl iron [10]. The mechanical properties can control just like the MRF did when external magnetic was applied. The advantage of MRE had overcome the critical issue in MRF, such as sealing problem, sedimentations, and environment contamination [11]. Additionally, it is expected that the flexibility in MRE material potentially overcome the limitation of the piezoelectric actuator. However, the research regarding of implementation MRE materials in the actuator is remain unclear. Therefore, the objectives of this paper to develop vision closed loop control system for MRE linear actuator including the plant model and PID controller. The signal response was extracted from Arduino serial plotter and replot using Microsoft Excel. The data is needed in the system identification for development of the plant model closed loop control system using MATLAB.

2. Methodology

Figure 1 shows the complete testing setup developed for linear MRE composite actuator using vision-based positioning system.

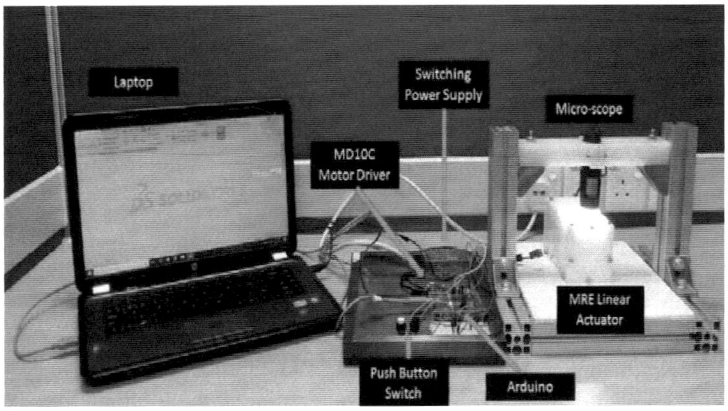

Figure 1. Experimental setup

3. Development of closed loop control system

3.1 Dynamic model of MRE linear actuator

The dynamic model of the linear actuator based on MRE was constructed, and the dynamics response was observed. Before that, the step input was converted to sine waves using the Fourier Transform, as shown in figure 2. The step input was constructed using Arduino in order to produce the displacement response of the MRE linear actuator. Besides, the response time and PWM was set to 200-time frame (8 s) and 250, respectively. This step input was embedded into MD10C motor driver, and then displacement of the MRE actuator was generated.

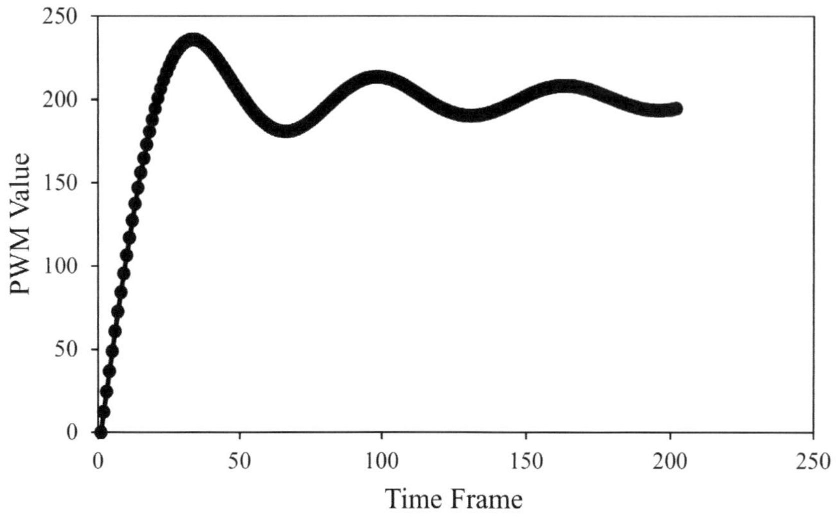

Figure 2. Sine wave input signals

Meanwhile, figure 3 shows the displacement response of the MRE linear actuator without embedded in the closed-loop control system for different current. This response was recorded using the vision-based positioning system. The graph shows that the response time for steady-state displacement was 5 seconds for 0.34 A, 0.37 A and 4.0 A of currents, respectively. Since the trend of displacement responses were similar to the sine wave input signal regardless of the current. Therefore, this displacement response could be used as an output response in developing the plant model using the system identification tool.

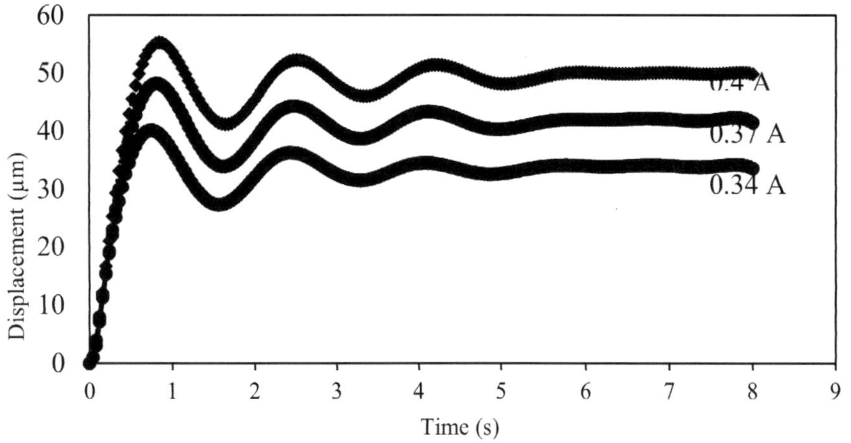

Figure 3. Displacement responses for 0.34 A, 0.37 A and 4.0 A of current

3.2 Plant model for MRE Linear actuator.

A plant is the combination of the process and actuator and commonly referred to with a transfer function. This transfer function was describing the relation between input (sine waves) and output (displacement response) signal of the system. In this work, the transfer function for MRE linear actuator was constructed using system identification tool which available in the MATLAB version R2016b.

Table 1 indicates that the fit percentage for transfer function by varying the number of poles and zeros. The highest fit percentage was 88.56% using 3 and 2 of the number of poles and zero, respectively. This is followed by 80.14% using 3 and 1 the number of poles and zero, respectively. However, the lowest fit percentage measured was 51.59% using 2 and 1 of the number of poles and zero, respectively. The plant models based on the fit percentage and input model (sine waves) was plotted, as shown in figure 4.

Table 1. Fitted parameter.

Zero	Poles	Best fit (%)
1	2	51.59
1	3	80.14
2	3	88.56

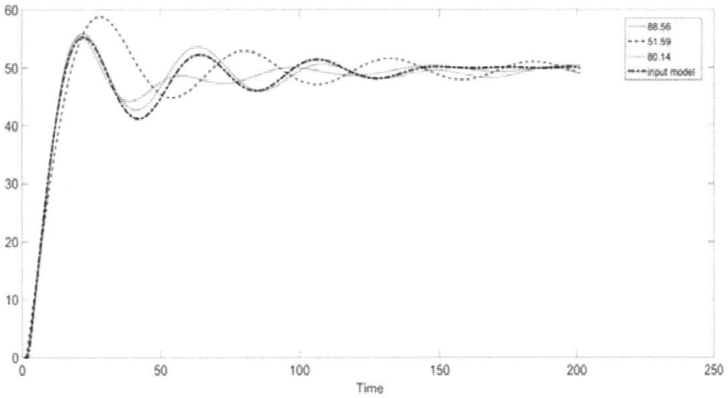

Figure 4. Plant model plots

A mathematical equation was developed to describe the plant model for MRE linear actuator. The degree of the numerator of the mathematical equation indicates the number of zero and the degree dominator indicate the number of poles. Since then, the degree of numerator cannot exceed the degree of the dominator for proper plant model. The equation (1), (2) and (3) represented the mathematical equation of 88.56%, 80.14% and 51.59% fit percentage, respectively. However, the stability of the system unable to describe by using this mathematical equation.

$$\frac{0.0958\, s^2 + 0.003642\, \mathrm{s} + 0.001549}{s^3 + 0.3572\, s^2 + 0.2805\, s + 0.006333} \tag{1}$$

$$\frac{0.0005982\, \mathrm{s} + 0.0001467}{s^3 + 0.1346\, s^2 + 0.2892\, s + 0.0005872} \tag{2}$$

$$\frac{0.4123\,s + 0.03714}{s^2 + 1.753\,s + 0.1499} \quad (3)$$

Thus, the number of the pole (x) and zero (o) was plotted, as shown in figure 3.4 in order to observe the stability of the system. The stability of the system was determined to identify the position of the pole and zero in the plot. The system was considered unstable when the plot was in the right half of 's' plane. Since all the plot was in the left half of 's' plane, the system was considered stable. Therefore, 88.56% of the best fit was considered as a plant model for MRE linear actuator, which will use in the evaluation of the closed-loop control system performances.

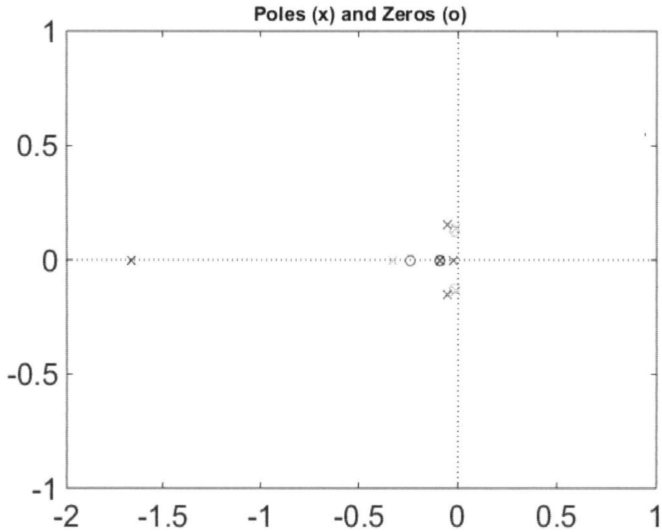

Figure 5. Pole and Zero plot

3.3 PID controller

From the best-fitted plant model obtained from previous work, the Proportional-Integral-Derivative controller (PID) was developed to be used as a controller in the closed-loop control system. Therefore, the PID controller was develop using PID tuner in the MATLAB. In the PID tuner, the transfer function was imported to the plant section. The parallel, and time for form and domain, respectively. The PID was tuned until the optimum settling time. The simulation PID controller and without PID controller will be discussed in next sub chapter.

3.4 PID tuner

From the previous work, the constructed plant model of MRE linear actuator was exported to the PID tuner in order to develop the robustness of the closed-loop control system. The PID value was embedded in the plant model to improve the time response and reduce the percentage of the overshoot error in the model. The mathematical equation of the transfer function with PID was described in equation 3.4. This shows that the model of MRE linear actuator was precisely developed and applicable for the actual system.

$$\frac{0.5083s^2 + 0.01932s + 0.008221}{s^3 + 0.8655\, s^2 + 0.04737s + 0.01455} \quad (4)$$

In figure 6, the transfer function model with and without PID controller was compared in order to observe the amplitude and settling time. The settling time for the PID controller was faster than without a PID controller with the value of 7.87 s and 16.9 s, respectively. Besides that, the amplitude of the PID controller also higher than without the PID controller with the value of 0.6006 and 0.275, respectively. Table 3.3 shows the summary performances of the plant model with and without PID controller. The PID controller has reduced the overshoot response from 12.5 to 7.01. Furthermore, the rise time for the PID system was faster with the value 0.201 s compared to the without PID system, the rise time was 0.391 s.

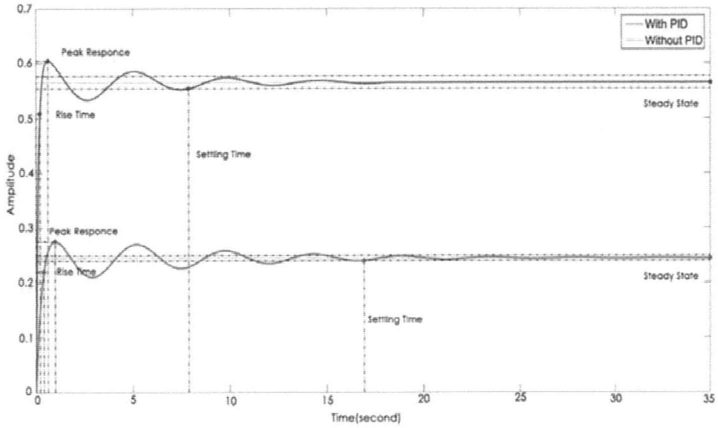

Figure 6. PID simulation

Table 2. Performances of the plant model

		With PID	Without PID
Settling time (s)		7.87	16.9
Peak Response	Peak Amplitude	0.6006	0.275
	Overshoot	7.01	12.5
	Peak Time (s)	0.68	1
Rise Time (s)		0.201	0.391
Steady State		0.565	0.245

According to the results above, the speed of the simulated PID system considers as moderate due to the rise time. Therefore, this control design with PID was possible to be used in the closed-loop control system based on the MRE linear actuator. This is because the high speed will be caused by positioning errors due to the elastic damping but the speed cannot be too low since it is not suitable to be used in the application. Lastly, the closed-loop control system for MRE linear actuator was developed using script code in MATLAB. Figure 7 shows the schematic diagram of the complete process. The algorithm of the closed-loop control system was described in Appendix B.

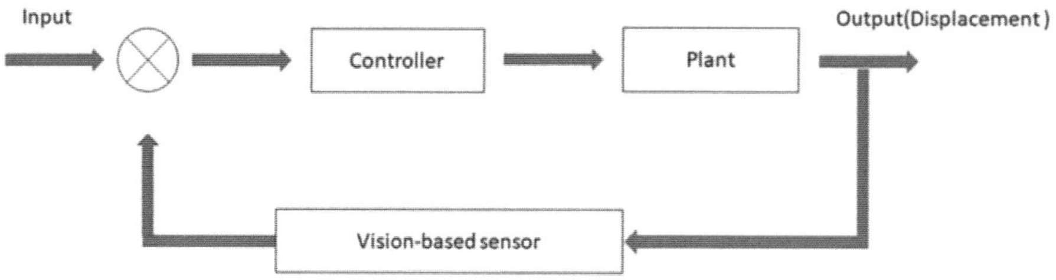

Figure 7. Closed-loop Control System Diagram

3.5 Closed – loop control system

The closed-loop control system was developed to control the displacement response for MRE linear actuator. This is to test the capability of the PID controller to communicate with the MRE linear actuator in order to produce the desired positioning.

In the evaluation of closed-loop control system for MRE linear actuator, the input displacement magnitude was considered as manipulated input. Figure 8 shows the displacement response of actuator for 10, 20, 30, 40, and 50 µm of displacement magnitude applied to the closed-loop control system.

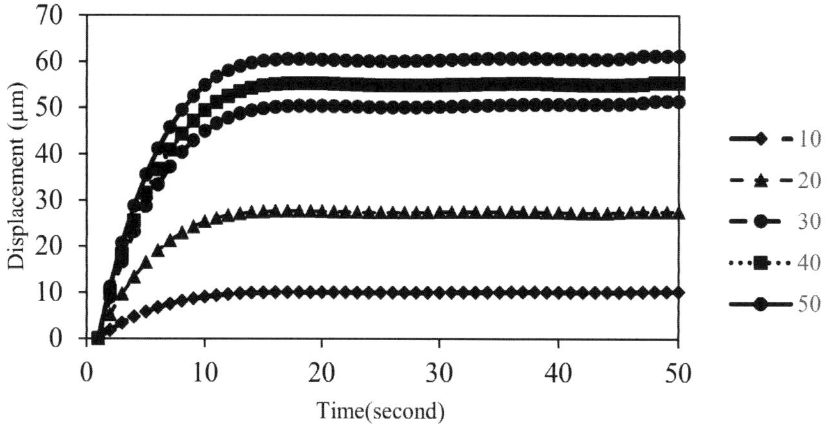

Figure 8. Displacement response for difference displacement.

The graph shows that displacement response of the MRE linear actuator was increased from 10 µm to 60 µm. However, the settling time was similar with differences input displacement. The displacement rate for the closed-loop control system achieved steady-state was 1.05, 2.61 and 4.63 µm/s, for 10, 20 and 30 µm, respectively. However, the further increment of the input displacement magnitude on the closed-loop control system, the displacement rate remained constant, which was 5.08 µm/s, as shown in figure 3.8.

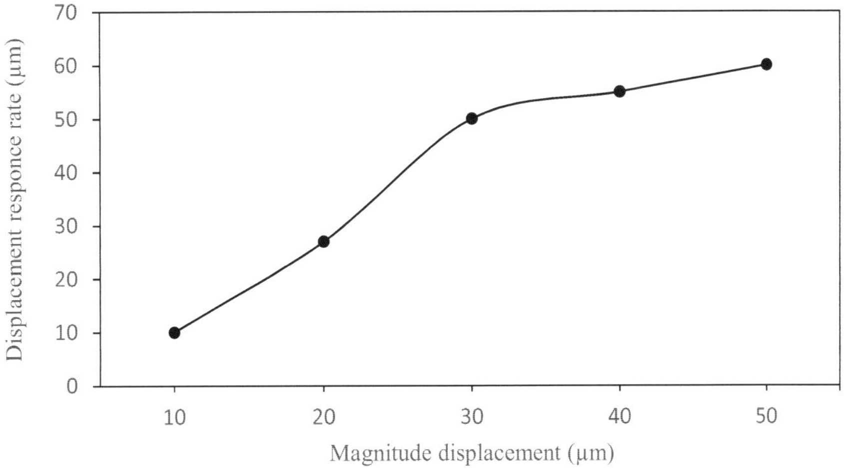

Figure 9. The displacement response rate for 10, 20, 30, 40, and 50 μm of input magnitude displacement.

On the other hand, it is expected that the displacement response from the control system was similar to the input displacement magnitude. The closed-loop control system accurately generated the 10 μm of the input displacement magnitude. However, there was an error for 20 μm, 30 μm, 40 μm and 50 μm. The highest steady-state errors were 67% obtained from the input displacement of 30 μm. It is followed by 38%, 35%, and 20% for 40 μm, 20 μm and 50 μm, respectively. The steady-state errors were closely related to the magnetic hysteresis of the magnetic materials, which was MRE. Furthermore, the steady-state error increased from 10 μm to 30 μm and then this value was reduced, as shown in figure 3.9. This is because the magnetic density was close to the saturation stage at 60 μm.

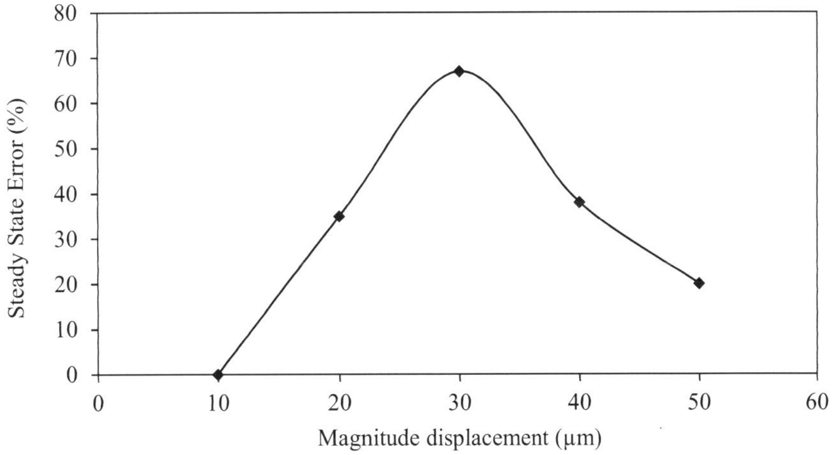

Figure 10. The steady-state error rate for 10, 20, 30, 40, and 50 μm of input magnitude displacement.

Lastly, the closed-loop control system was successfully developed for MRE linear actuator, and the performances of the closed-loop control system were summarized in table 3.4. The displacement response was constructed as the input magnitude displacement with steady-state errors. However, the developed system was considered acceptable.

Table 3. Summary of the closed-loop control system performances.

Displacement (μm)		Settling time (s)	Displacement response rate (μm/s)	Steady State Error (%)
Input	Output			
10	10	12	1.05	0
20	27	12	2.61	35
30	50	12	4.63	67
40	55	12	5.08	38
50	60	12	5.08	20

4. Conclusion
A plant model for the MRE linear actuator was successfully developed using system identification approach. The developed model showed excellent stability with the best-fit percentage of 85.56% from the input response. The system with a PID controller has significantly improved the settling time and amplitude response. The speed of the simulated PID system was moderate, which is suitable to be used in the closed-loop control system for MRE linear actuator. Lastly, the closed-loop control system was successfully developed for the MRE linear actuator. The system was tested with a different magnitude displacement input, and the control system was capable of generating the displacement response with a certain amount of steady-state error.

5. References
[1] G. A. Dominguez, M. Kamezaki, M. French, and S. Sugano, "Modelling and simulation of a new magnetorheological linear device," pp. 235–240, 2016.
[2] M. Rakotondrabe, *Smart Materials- Based Actuators at the Micro / Nano-Scale*. 2013.
[3] J. Mohd Jani, M. Leary, A. Subic, and M. A. Gibson, "A review of shape memory alloy research, applications and opportunities," *Mater. Des.*, vol. 56, pp. 1078–1113, 2014.
[4] J. L. Zhong, "Smart Materials and Nanotechnology in Engineering, SMNE2011," vol. 345, 2012.
[5] I. Grinberg, N. Maccabi, A. Kassie, and D. Elata, "A Piezoelectric Twisting Beam Actuator," *J. Microelectromechanical Syst.*, pp. 1–8, 2017.
[6] J. Guo, S. K. Chee, T. Yano, and T. Higuchi, "Micro-vibration stage using piezo actuators," *Sensors Actuators, A Phys.*, vol. 194, pp. 119–127, 2013.
[7] F. Gordaninejad, X. Wang, and P. Mysore, "Behavior of thick magnetorheological elastomers," *J. Intell. Mater. Syst. Struct.*, vol. 23, no. 9, pp. 1033–1039, 2012.
[8] N. M. Wereley, *Magnetorheology*. 2013.
[9] M. Ashtiani, S. H. Hashemabadi, and A. Ghaffari, "A review on the magnetorheological fluid preparation and stabilization," *J. Magn. Magn. Mater.*, vol. 374, pp. 711–715, 2015.
[10] M. Bunoiu and I. Bica, "Magnetorheological elastomer based on silicone rubber, carbonyl iron and Rochelle salt: Effects of alternating electric and static magnetic fields intensities," *J. Ind. Eng. Chem.*, vol. 37, pp. 312–318, 2016.
[11] Y. Li, J. Li, W. Li, and H. Du, "A state-of-the-art review on magnetorheological elastomer devices," *Smart Mater. Struct.*, vol. 23, no. 12, p. 123001, 2014.

Acknowledgement
This research is supported in part by Mybrain15 and Universiti Kuala Lumpur for financial support. In addition, the authors like to express appreciation to the System Engineering and Energy laboratory member and Universiti Kuala Lumpur Malaysia Spanish institute technical staff for the encouragement and advice is gratefully acknowledged.

IOP Conf. Series: Materials Science and Engineering **815** (2020) 012019 doi:10.1088/1757-899X/815/1/012019

Numerical Study of the Flow-Induced Vibration During Tunnel Surging at the Intake Section of Chenderoh Dam

AZA Mazlan[1], MH Zawawi [2*], A Arbain[1], MHM Ghazali[1], MHA Satar[1], MA Abas[1], NH Hassan[2] and MRM Radzi[3]

[1]School of Mechanical Engineering, Engineering Campus, Universiti Sains Malaysia, 14300 Nibong Tebal, Pulau Pinang, Malaysia
[2]Uniten R&D Sdn Bhd (URND), Universiti Tenaga Nasional, 43000 Kajang, Selangor, Malaysia
[3]TNB Generation Division, Tenaga Nasional Berhad, 59200 Kuala Lumpur, Malaysia

*Corresponding author: MHafiz@uniten.edu.my

Abstract. Chenderoh Dam that located in Malaysia is one of renewable energy power plant that beneficial to mankind. However, in some cases, the dam suffers from the vibration effect during the water spilling from upstream to downstream. This study focused on major part of the dam which is the intake section during the tunnel surging condition. A detail 3D model of the intake section was constructed and used in the prediction of flow-induced vibration response. The results of frequency domain response and operational defection shapes (ODS) from the effect of flow-induced vibration are compared with the natural frequencies and mode shapes of the dam. From the results, the transient vibration responses due to the flow of water occurred at the frequency of 8.63 Hz with the maximum deformation of 8.24 x 10^{-1} m, meanwhile, the modal analysis obtained at 2.76 Hz of natural frequency with deformation of 9.1 x 10^{-4} m. The deformation of ODS is high because of the water flow and tunnel surging condition. However, there is no resonance phenomenon occurred, yet a safety precaution must still be considered by the operator based on this result.

1. Introduction

A major interest in dam study is the prediction using fluid-structure interaction (FSI) which based on the finite volume method (FVM). Both interactions can be described as stable or oscillatory, between the structure and surrounded fluid-flow. To date, many researchers used this approach as prediction in determining the flow-induced vibration studies [1]. For examples, the studies on the effect of sector gate[2], radial gate [3,4], sliding gate [5] and double curvature arch dam [6] to the dam structure and water flow have been conducted. Another study on the effect of reservoir water level fluctuation to the non-linear seismic behavior of high concrete dam also has been done [7].

Vibration study on the dam structure offers different prospective learning, whereas a safety measure can be taken where its considering people life. Generally, the vibration caused by the fluid-flow is a major factor in degradation and failure of the dam structure. For example, there is a study focused in finding an optimal control of valve for the dam operation. This study suggested that valves with rounded edges and a sloping bottom has lower deflection amplitude which can reduced the vibration force and less possibility to damage the dam wall [5]. Fei Chong et. al. studied the effect of different radial gate height with force exerted to the wall during water flow. The findings show that, the 8 m gate height exerted lower force which can reduced the vibration effect [4]. Another significant finding on vibration induced by fluid is it can lead to the catastrophic failure if natural frequencies coincide with shedding frequencies of the flow. A good dam structure will have different trend of frequency response function (FRF) from the shedding vibration [8].

Operating deflection shape (ODS) is introduced to get full-field dynamic response of the structure. This method can describe the vibration pattern of any structure with influence by its known or unknown operating frequencies [9]. Some suggested that, ODS provides more reliable dynamic information than a mode shape since it can be produced at any single harmonic frequency and not restricted to the natural frequencies [10]. This is the main reason why ODS has been applied in variety damage identification such as in polymeric composite plates [10] and beam with different types of damage (i.e. surface slot, surface holes and fatigue cracks) [12].

Chenderoh Dam is an oldest dam in Perak region that located in Malaysia. None of numerical study has been conducted to study the vibration that related to the dam structure. Based on literatures, the finite element method (FEM) and FVM were chosen to predict the vibration characteristic specifically in crucial part which is intake section where the electricity is generated. The study will compare with ODS and modal analysis (MA) results to determine the deformation occurred and relate it with resonance phenomenon for the safety concern of the dam structure.

2. Methodology

2.1. Modelling and Boundary Condition

Intake section consists of three head gates, turbines and wicket gates. In this study, the case of head gates fully opened and wicket gate fully closed will be conducted as shown in Figure 1-4. So, the water will flow from upstream to downstream with maximum velocity and stopped by the wicket gates. The study only focused on the water surging effect when the wicket gates are closing inside the pen stock and the effect of the flow-induced vibration to the intake section. The real scale 3D model of intake section is constructed using SolidWorks software as shown in Figure 1 and 2.

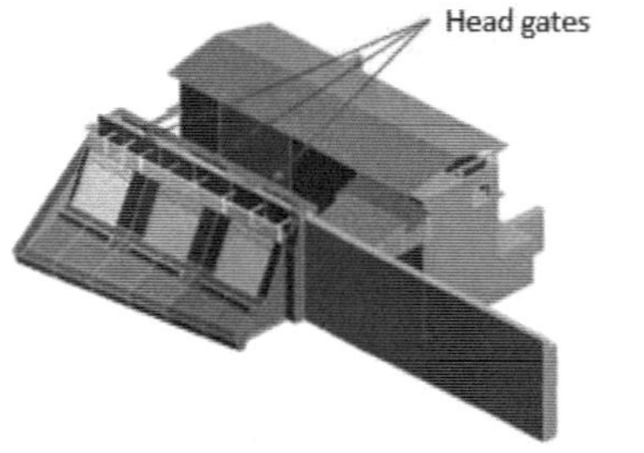

Figure 1. Isometric view of intake section

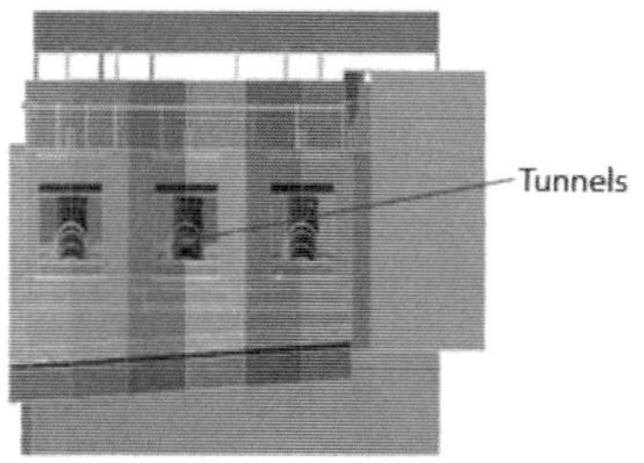

Figure 2. Back view of intake section

Figure 3. Fully opened wicket gate

Figure 4. Fully closed wicket gate

IOP Conf. Series: Materials Science and Engineering **815** (2020) 012019 doi:10.1088/1757-899X/815/1/012019

2.2. Modal Analysis (MA)

This analysis is conducted to determine the vibration characteristic such as resonant frequencies and mode shape [13]. Each part (i.e. penstock, wall and piping) were applied with its real material properties (i.e. concrete, steel and zinc) from the engineering data. The geometry is meshed with tetrahedra element for better accuracy [14]. Figure 5 shows meshes that represented the physical domains of the geometry. After that, the base (blue faces) of the structure is set as fixed support to represent the foundation and restrict its degree of freedom as shown in Figure 6.

Figure 5. Overall meshing on the geometry

Figure 6. Fixed support (Base)

Then, the modal analysis is performed to determine the dynamic properties at certain range of frequency called FRF. The dynamic properties can be simulated in terms of deformation, velocity or acceleration [15]. In the analysis setting, the gravity force with magnitude of 9.81 ms^{-1} is applied to all three directions. The solution method used is mode superposition interval and set to 50 natural frequencies and mode shapes.

2.3. Fluid Structure Interaction (FSI)

This analysis is used to determine the interaction of the dam structure with fluid region from upstream to downstream. The type of mesh and boundary condition used is the same as in MA. In the analysis setting, the time step end is set to 5 seconds and maximum-minimum sub-step are set to 1 and 100 respectively. The earth gravitational acceleration is applied to negative y-direction with magnitude of 9.81 ms^{-1}. In the initial fluent setting, the fluid is modelled as 3D incompressible flow and the water-air is modelled using implicit VOF. Figure 7 and 8 shows the setting of inlet and outlet for the intake section, respectively. The intake part is supplied with water velocity of 1 ms^{-1}. For the interaction of fluid and structure, the system coupling is executed to analyse both data transfers. In this system, the end time is set to transient structural system which is 5 seconds and vice versa for time step size.

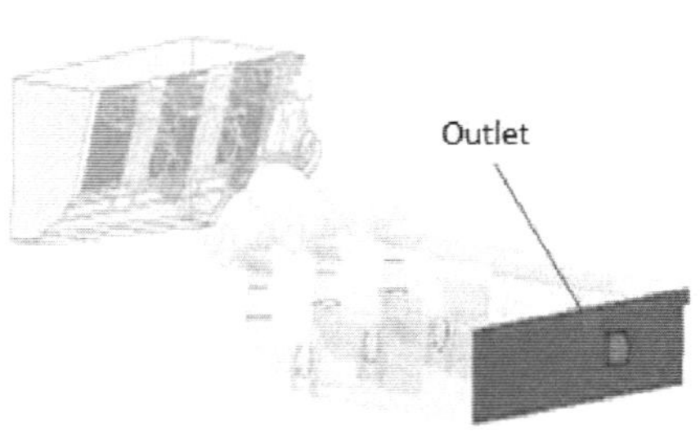

Figure 7. Inlet section for intake part

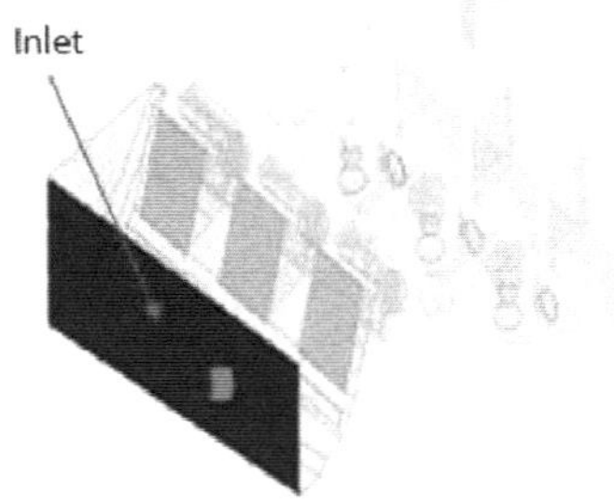

Figure 8. Outlet section for intake part

IOP Conf. Series: Materials Science and Engineering **815** (2020) 012019 doi:10.1088/1757-899X/815/1/012019

The deformation data obtained from FSI analysis is then converted to frequency domain data through the fast Fourier transform (FFT) equation. After that, the FFT and FRF graph are observed and compared.

3. Results and Discussion

3.1. MA Results of Intake Section

From the analysis, the mode shape corresponds to the flow of water towards the downstream area with the amplitude value of 9.1 x 10^{-4} m and the natural frequency of 2.76 Hz. This value of 2.76 Hx must be taken into consideration to avoid resonance phenomenon. Figure 9 shows the mode shape of the intake section that affected the penstock and roof areas of the intake section. From the harmonic response analysis, the highest deflection occurred at z-direction with amplitude value of 7.27 x 10^{-10} m, as shown in figure 10. The highest peak of the graph represents the natural frequency value of 2.76 Hz.

Figure 9. Mode shape of the intake section at the natural frequency of 2.76 Hz

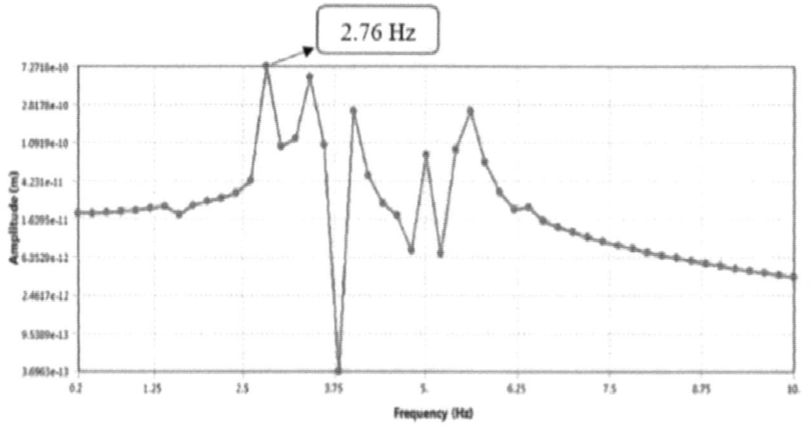

Figure 10. FRF graph in z-direction of the intake section

3.2. ODS Results of Intake Section

Figure 11(a) shows the ODS of the intake section in time range of 5 seconds. The maximum displacement occurred at trash rack with overall displacement of 8.24 x 10^{-1} m (average). In time domain, the deformation is likely to decay along time as shown in Figure 11(b). At $0 < t < 1$, the water surging from upstream to downstream cause the transient deformation at the intake structure. For $t > 1$, the water flow reached steady-state and deformation decrease along the time.

IOP Conf. Series: Materials Science and Engineering **815** (2020) 012019 doi:10.1088/1757-899X/815/1/012019

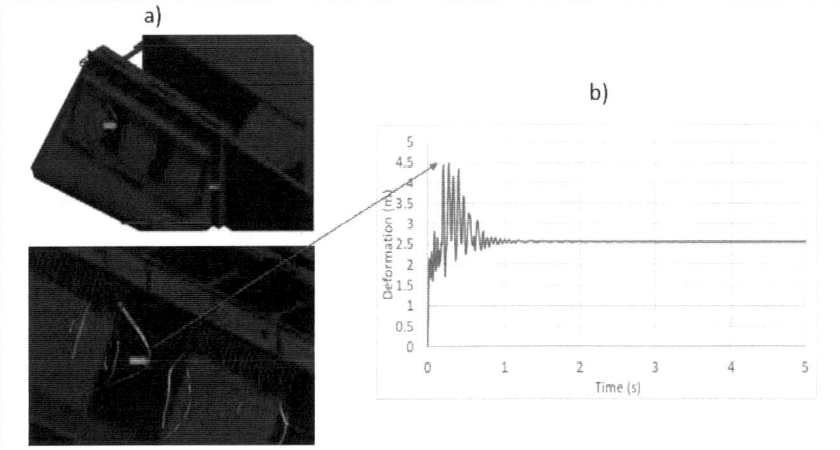

Figure 11. (a) ODS of intake section and (b) time domain of deformation

3.3. Comparison between MA and ODS

Figure 12 and 13 show the deformation occurred in the intake section from the MA and ODS results. The highest deformation found at the trash rack location with the deformation value of 8.24 x 10^{-1} m. This indicate that this area more likely to fail due to high velocity of water and sudden closing of the wicket gate. Differently for the MA result, the deformation occurred at rooftop with an amplitude of $9.1 \times 10^{-4}\,m$. For operating and natural frequencies, it was induced at 8.63 Hz and 2.76 Hz, respectively. In overall, the transient vibration effect with operating frequency of 8.63 Hz is not coincides with the natural frequency of MA at 2.67 Hz, thus there will be no resonance phenomenon for this case. The first peak of ODS occurred due to the first impact of water spilling with wicket gate that close suddenly. Afterwards, the deformation tends to reduce based on the residue of the first transient effect of the surging. The trend for ODS is likely diverge with modal analysis. Unfortunately, due to water surging effect, the deformation is high and may cause damage the to the structure especially on trash rack. A safety concern should be considered as precaution where the wicket gate should be close properly to prevent any damage on trash rack part.

Figure 12. The deformation prediction using MA

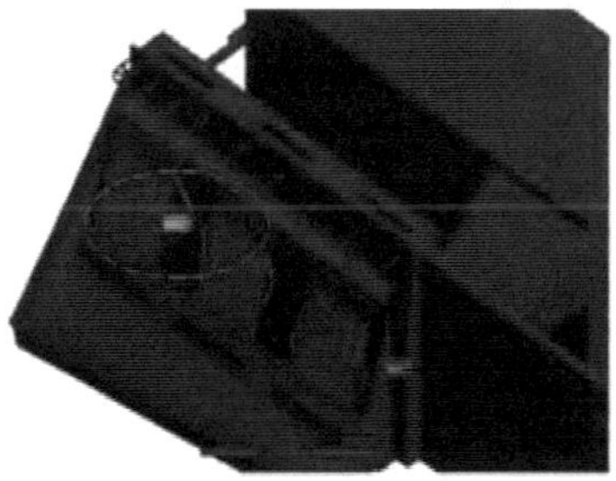

Figure 13. The deformation prediction using ODS

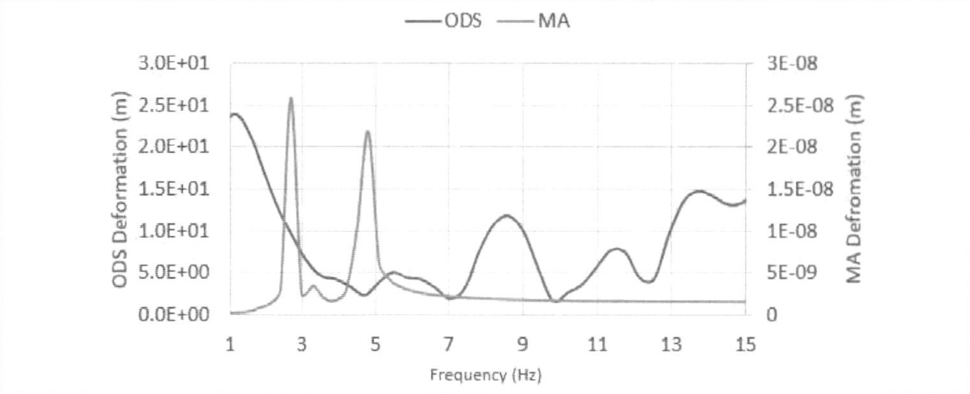

Figure 14. Comparison between MA and ODS deformation in frequency domain

4. Conclusion

Based on this study, the foresight of the vibration response at the intake section of the Chenderoh Dam can be concluded as follows:

1. A numerical method was introduced in this study to model and simulate the real scale of intake section of the Chenderoh Dam. Using the transient structural, fluent and system coupling in ANSYS, the two-way FSI has been completed in 5 second of time.
2. From ODS results, the highest deflection occurred at trash rack with operating frequency of 8.63 Hz and induced displacement of 8.24×10^{-1} m.
3. Differently for the MA, the highest deflection occurred at rooftop with natural frequency of 2.6 Hz and exhibit displacement only 9.1×10^{-4} m.
4. For comparison of both results, the trend is diverged to each other in term of frequencies and patterns. However, the deformation occurred in transient analysis (ODS) is high, so precaution steps should be taken to prevent any damage to the dam structure.

5. References

[1] M. H. M. Ghazali *et al.*, "Prediction of the Flow-Induced Vibration Response of the Chenderoh Dam Left Bank Section," *MATEC Web Conf.*, vol. 217, pp. 1–6, 2018.

[2] M. H. M. Ghazali *et al.*, "Structural Dynamic Analysis of the Chenderoh Dam Sector Gate Section," *MATEC Web Conf.*, vol. 217, pp. 1–6, 2018.

[3] S. O. Lee, H. Seong, and J. W. Kang 2018 Flow-induced vibration of a radial gate at various opening heights *Eng. Appl. Comput. Fluid Mech.* **12** no. 1 pp. 567–583

[4] F. C. Ng et al. 2018 Fluid structure interaction study on the variation of radial gate's gap height in dam *IOP Conf. Ser. Mater. Sci. Eng.* **370** no. 1

[5] I. Javanshir, N. Javanshir, R. Barmaki, and M. Mahmoodi-k 2015 Modelling of the fluid-induced vibrations in sliding gate dams *JVE Int. Ltd. J. of Vibroeng.* **17** no. 1 pp. 478–486

[6] R. Manikandan and P. Sulthana 2014 Fluid Structure Interaction (FSI) of Double Curvature Arch Dam Under Seismic Loading By the Application of Added Mass Technique and Acoustic Elements *COMSOL Conf.*

[7] M. A. Hariri-Ardebili and H. Mirzabozorg 2012 Reservoir Fluctuation Effects on Seismic Response of High Concrete Arch Dams Considering Material Nonlinearity *J. Civ. Eng. Res.* **1** no. 1 pp. 9–20

[8] W. Z. Lim and R. Y. Xiao 2015 A Comparison of the Fluid-Structure Interaction Method for Simple Numerical Analysis of Concrete Gravity-Dam

[9] S. S. Wang, Y. F. Zhang, M. S. Cao, and W. Xu 2015 Effects of Contraction Joints on Vibrational

Characteristics of Arch Dams: Experimental Study *Shock Vib.*

[10] A. Katunin and P. Przystałka 2014 Damage assessment in composite plates using fractional wavelet transform of modal shapes with optimized selection of spatial wavelets *Eng. Appl. Artif. Intell.* **30** pp. 73–85

[11] A. Katunin 2011 Damage identification in composite plates using two-dimensional B-spline wavelets *Mech. Sys. and Sig. Proc.* **25** no. 8 pp. 3153–3167

[12] P. F. Pai and L. G. Young 2001 Damage detection of beams using operational deflection shapes Int. *J. Solids Struct.* **38** no. 18 pp. 3161–3192

[13] M. Imran, A. A. Abbasi, and M. J. Hyder 2017 Determination of modal characteristics of cantilever beam *ICET 2016 - 2016 Int. Conf. Emerg. Technol.* **2** no. 2 pp. 4–6

[14] E. Wang, T. Nelson, and R. Rauch 2004 Back to Elements - Tetrahedra vs . Hexahedra *CAD-FEM GmbH.* p. 16

[15] B. J. Schwarz and M. H. Richardson 1999 Experimental Modal Analysis *CSI Reliab. Week* **35** no. 1 pp. 1–12

Acknowledgements

The authors acknowledge UNITEN R & D Sdn Bhd (URND) on the fund provided for this research under the TNB consultancy grant (U-TG-CR-18-04) and Universiti Sains Malaysia through RUI Grant 1001.PEMAKNIK.8014129.

Vibration Analysis of Chenderoh Dam Physical Model Spillway Structure due to the Effect of Water Spilling: Numerical and Experimental Validation

A Arbain[1], M H Zawawi[2], R Ganesan[1], M H M Ghazali[1], M R M Radzi[3], N H Hassan[2] and A Z A Mazlan[1]*

[1]School of Mechanical Engineering, Engineering Campus, Universiti Sains Malaysia, 14300 Nibong Tebal, Pulau Pinang, Malaysia
[2]Uniten R&D Sdn Bhd (URND), Universiti Tenaga Nasional, 43000 Kajang, Selangor, Malaysia
[3]TNB Generation Division, Tenaga Nasional Berhad, 59200 Kuala Lumpur, Malaysia

*Corresponding author: zhafran@usm.my

Abstract. Dam can be defined as a barrier built across the river to restrain and control the flow of water. However, the safety of the dam structure can be threatened by the uncertainties vibration induced from internal and external sources. Thus, in this paper, the vibration analysis in terms of Experimental Modal Analysis (EMA) and Operational Deflection Shapes (ODS) are carried out to determine the reliability of the dam structure. This study only focused on the spillway structure of the Chenderoh Dam that located in Perak, Malaysia. For the simulation, a 3D physical model with 1:20 scale is developed using SolidWorks software and simulated using ANSYS software. Whilst, for validation of the simulation results, EMA and ODS experiment are performed on the physical model of the Chenderoh Dam. The validation will be in terms of mode shape, ODS, natural and operating frequencies. From the results, the first natural frequency of the spillway occurred at 220.87 Hz with the maximum deflection of 1.6 mm and the ODS deformation happened at operating frequency of 45 Hz with the amplitude value of 0.003 m. The operating frequency value is far from the natural frequency of 220.87 Hz; thus, the transient vibration only induces a minimal effect on the spillway structure and the spillway is considered safe for the operation.

1. Introduction

The safety of the dam structure can be threatened by the uncertainties vibration induced from the internal and external sources. This includes the flow condition of water, earthquake effect, vibration of mechanical machine and other related sources. Various vibration studies and analyses have been conducted by many researchers. Such study includes eigenvalue analysis, normal condition analysis [1] as well as Experimental Modal Analysis (EMA) and Operational Deflection Shapes (ODS). Other than that, Yang [2] carried out the static and dynamic analyses to determine the maximum dynamic displacement and seismic stress of the dam. Besides, new formulas such as to calculate the frequency using an analytical method of the dam [3] and random vibration formula to analyse the dynamic characteristics of the dam [4] were introduced as well.

Simulation study is very crucial to be carried on in order to determine the vibration characteristics of the dam. Through this study, the dam structure behaviour under dynamic loading can be determined and thus can maintain the safety of the structure [5]. Sato & Obuchi [6] carried out simulation study to compare the dynamic response of the equivalent linear and non-linear finite element (FE) method. Modal analysis is another example of simulation to study the dynamic properties such as natural frequencies and mode shapes [7]. This is very important in determining, improvising and optimizing the dynamic characteristic of engineering structures [8]. There are two ways to conduct modal and harmonic analysis; using APDL or ANSYS Workbench. ANSYS Workbench is preferable as it is much easier to use compared to APDL. There are studies of modal analysis that been carried out using ANSYS Workbench to investigate and analyse and the dynamic behaviour of a structure [9, 10]. Beside simulation, there is another method to perform modal analysis which is by EMA [11].

EMA analysis is done to determine the modal parameters such as natural frequencies, mode shapes and damping values [12, 13]. For example, Balsara & Fowler [12] have conduct a study on two electromagnetic vibrators that used as exciter with a total of 14 points for velocity and 4 points for acceleration measurement for the tests. Meanwhile, a servohydraulic vibration generator was used as the exciter and 247 points were selected and measured [13]. ODS also can be used to study the vibrations of the structure. In ODS, the deflection shapes of a structure under different operating conditions can be determined. According to Hugar and Venkatesh [14], ODS is a vibration testing and analysis which has the ability to animate the deformation characteristics of a vibrating structures of systems. Generally, ODS contains the overall vibration of any forced motion for two or more DOFs points of the structure [14, 15]. Saravanan and Sekhar [15] stated that, ODS is different from mode shape as ODS can be used for non-linear and non-stationary structural motion, while mode shape can only be used for linear and stationary motion. In this study, the vibration of Chenderoh Dam physical model will be investigated in terms of EMA and ODS using both numerical and experimental.

2. Methodology

2.1. 3D modelling of spillway physical model

In this study, a spillway 3D model with 1:20 scale has been constructed using SolidWorks 2017 software. This spillway is mainly used for regulating the water from upstream to downstream area based on the opening distance between radial gate and sector gate. This section consists of the connection bridge, sector gate, radial gate and bridge to bottom outlet as shown in Figure 1. For the normal release of water condition, only radial gate is operated. However, when there is an emergency case, both radial and sector gates will be fully opened to release the water. It is expected that different gate openings will induce different vibration responses to the whole spillway section.

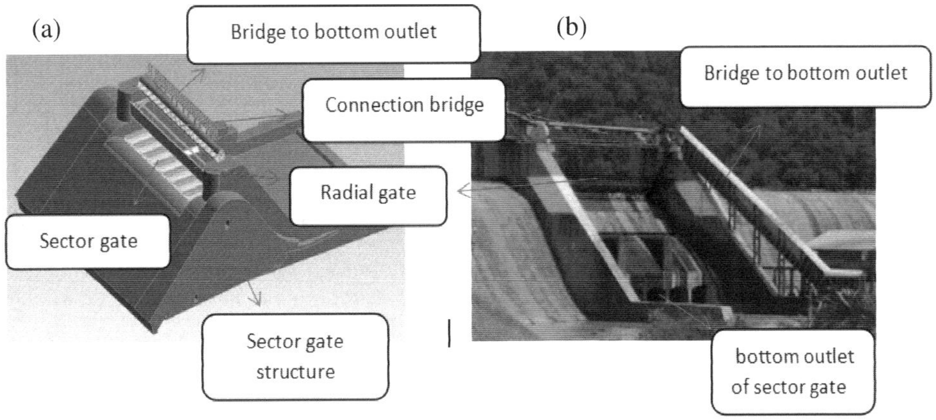

Figure 1. (a) 3D physical model of Chenderoh spillway section and (b) real photo of Chenderoh spillway section

2.2. Modal and harmonic response analyses of the spillway physical model

The modal analysis was carried out using ANSYS Workbench software. It is conducted to determine the vibration characteristics of the structure in terms of natural frequencies and mode shapes. Firstly, the 3D CAD model is imported from SolidWorks via Parasolid interface. Then, the desired materials were added along with the respective properties. In this structure, the material used is concrete and mild steel. The concrete with Young's modulus of E = 30 GPa is selected for the main spillway structure and mild steel (E = 210 GPa) is selected for other steel related parts. For the boundary condition, all bottom faces of the spillway structure are set as fixed support to represent the foundation of the dam structure as shown in Figure 2. A maximum of 30 mode shapes are set to be determined. An input force with magnitude of 9.81 N is applied in *x*, *y* and *z* directions. These forces act as assumption of gravitational forces that reacted in all three directions since the spillway section is simulated independently. The frequency range was set in accordance to the natural frequency obtained in modal analysis. The solution interval was set to 100.

Figure 2. Boundary condition of spillway (Fixed support)

2.3. Fluid structure interaction (FSI) analysis of the spillway physical model

For the FSI simulation study, there are three main elements needed. Firstly, the transient structural which used to study the deflection of the structure when fluid flow is act in the dam structure. Secondly, is the fluent where the flow and type of fluids that going to act on the dam structure. Thirdly, the system coupling where it combines these two, transient structural and fluent together. This method is called as two-way FSI in the ANSYS software, which is more accurate and stable solution but in the other hand, the one-way method has lower computational cost.

For a complicated structural design in FSI study, an appropriate type of mesh that can be used is tetrahedral mesh due to faster simulation time and much easier to be applied. For the analysis setting, the step end time is set to 5 second and the step control is changed from time to sub steps. The initial sub step, minimum sub step and maximum sub step is set as 1, 1 and 500 consequently. Boundary condition for the transient analysis is similar as in modal analysis where all the bottom faces is fixed as foundation and the standard earth gravity is set as 9.81 m/s^2. In fluent analysis, the type of boundary system is set as velocity inlet and pressure outlet as shown in Figure 3 and 4, respectively. The velocity magnitude of fluid is set to 1 m/s^2 and the fluid is modelled as incompressible flow with pressure-based solver.

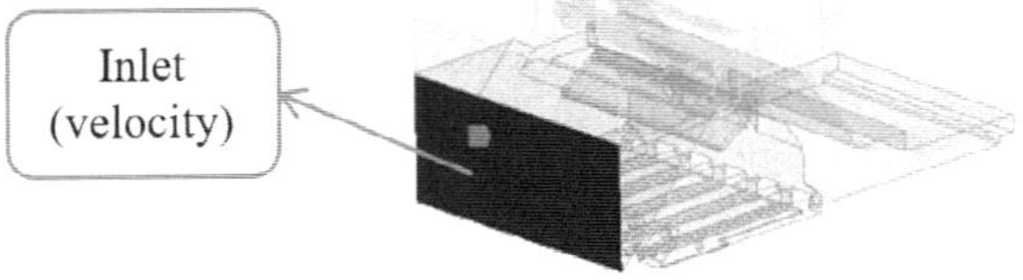

Figure 3. Inlet section for spillway part

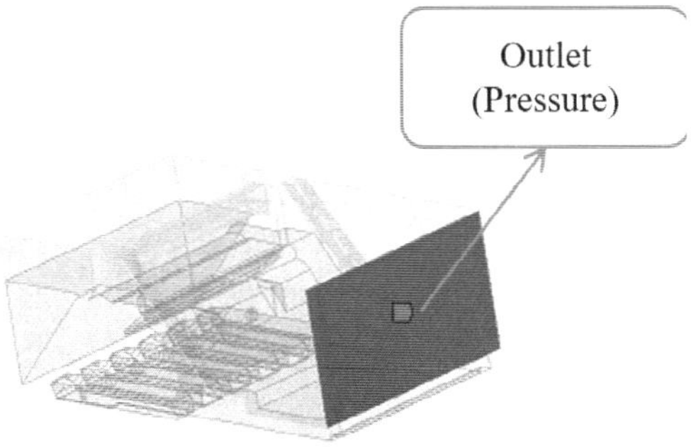

Figure 4. Outlet section for spillway part

2.4. Experimental validation of the physical model

The experimental modal analysis (EMA) is carried out using LMS Test Lab software and the result obtained was then compared with the ANSYS simulation. Both results are compared in terms of natural frequencies and mode shapes. The setup of EMA is shown in Figure 5. For EMA, an impact hammer is used to provide the input force to the structure and accelerometer is mounted to the respective node. The function of accelerometer is to measure the acceleration introduced from the structure vibration. To obtained best result, higher number of impact points are needed. Moreover, the impact hammer is knock at the same point at least 3 times to obtain the average reading for a better accuracy result. Apart from EMA, ODS setup is almost identical except it operates with different software. Consequently, ODS measurement is perform to study fluid excitation on the structure. Therefore, several accelerometers are attached at the nodes to get the operating frequency data and deflection shapes. Figure 6 shows the setup for ODS testing. Based on the figure, the water is flowing through the passage of physical model while the test is running.

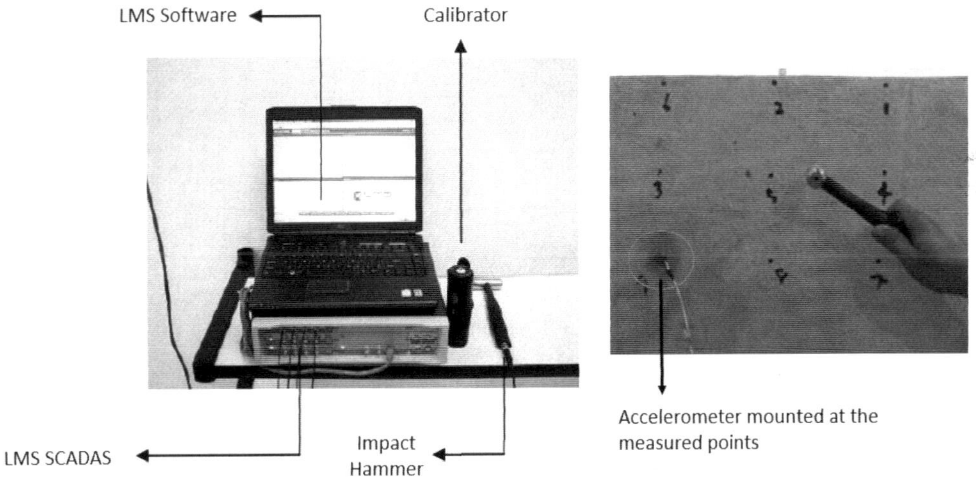

Figure 5. Experiment setup of EMA

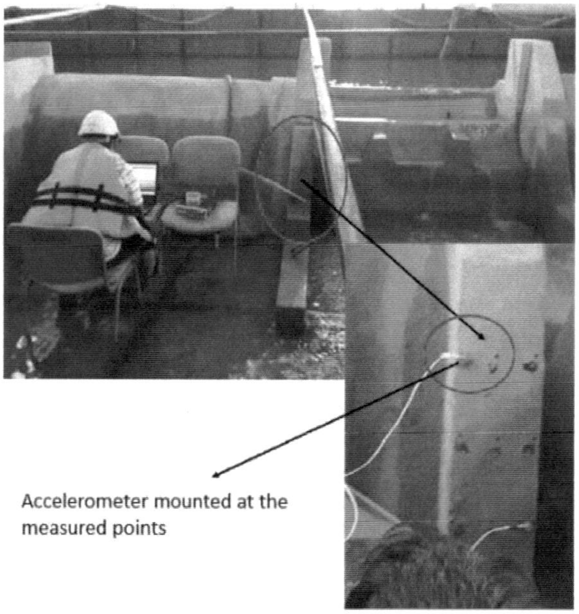

Figure 6. Experiment setup of ODS

3. Results and Discussion

3.1. Modal and harmonic analyses of the spillway physical model

Table 1 shows the six most significant natural frequencies and mode shapes of the spillway physical model. The 1st mode take place at the natural frequency of 220.87 Hz with the maximum deflection value of 1.6 mm. The 4th ,5th ,8th ,11th and 17th modes occurred at 397.83 Hz, 424.18 Hz, 484.64 Hz, 593.2 Hz and 705.61 Hz, respectively, with the highest deflection occurred at 4th mode with 3.9 mm. The red colour zone on the spillway structure indicates the highest or maximum deflection of the part.

Table 1. First six most significant frequencies and the mode shapes.

Modes No.	Mode Shapes	Natural Frequency	Deflection Values
1	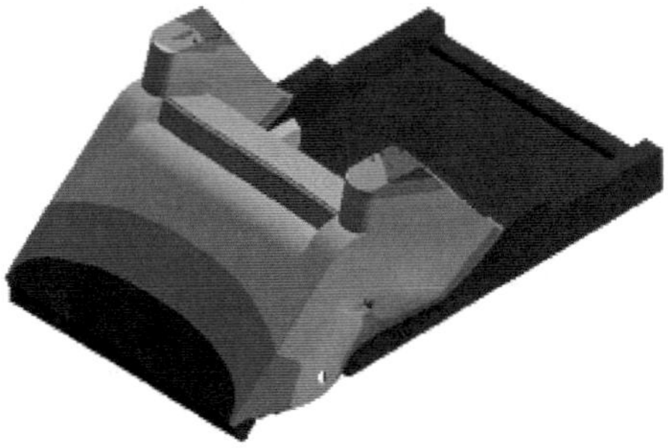	220.87 Hz	1.6 mm

IOP Conf. Series: Materials Science and Engineering **815** (2020) 012020 doi:10.1088/1757-899X/815/1/012020

4		397.83 Hz	3.9 mm
5		424.18 Hz	1.4 mm
8	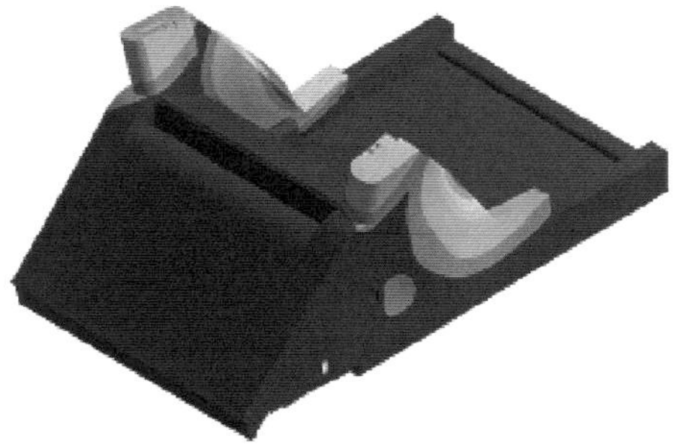	484.64 Hz	3.8 mm
11	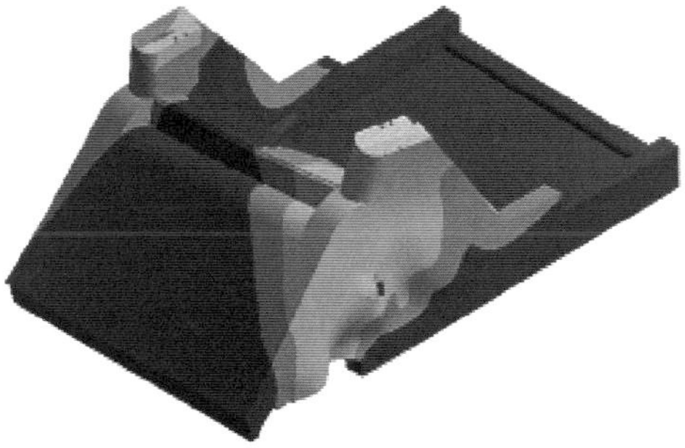	593.2 Hz	3.2 mm
17	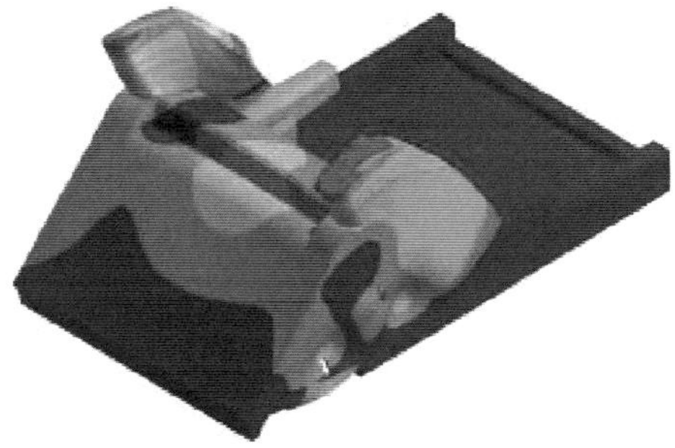	705.61 Hz	2.5 mm

IOP Conf. Series: Materials Science and Engineering **815** (2020) 012020 doi:10.1088/1757-899X/815/1/012020

Table 2 shows the FRF of the spillway in *x*, *y* and *z* axes directions. The highest natural frequencies for each direction occurred at 220 Hz, 620 Hz and 420 Hz, respectively with the maximum deflection value of 1.056 x 10^{-4} mm, 4.75 x 10^{-5} mm and 1.35 x 10^{-5} mm. The most significant vibration occurred in *x* axis direction which 220 Hz of natural frequency and deflection of 1.056 x 10^{-4} mm.

Table 2. FRF graph of spillway in *x*, *y* and *z* axis.

<table>
<tr><th>Axis</th><th>FRF Graph</th></tr>
<tr><td>x</td><td>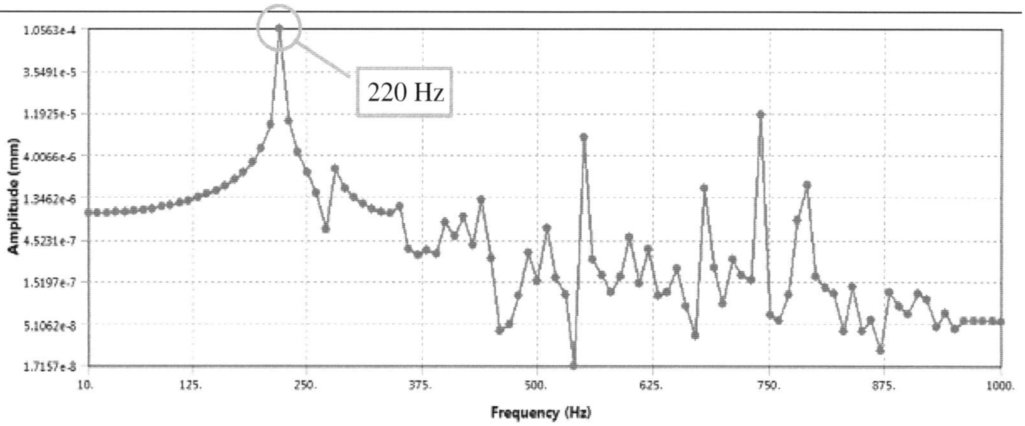
</td></tr>
<tr><td>y</td><td>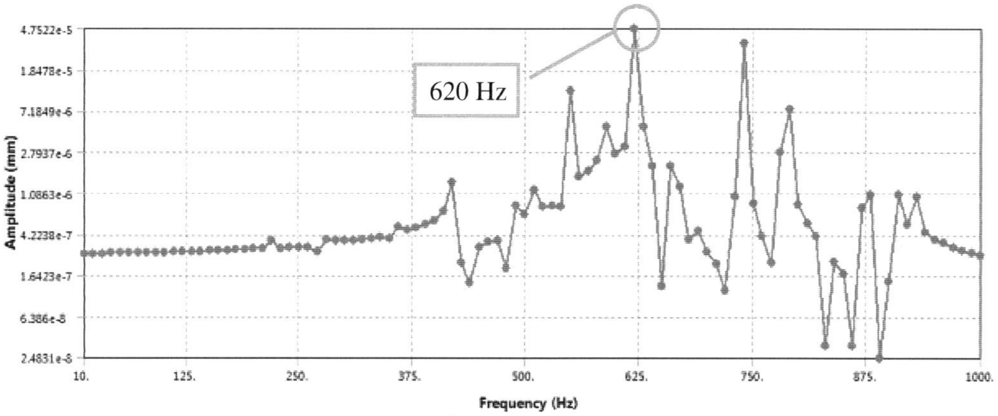
</td></tr>
<tr><td>z</td><td>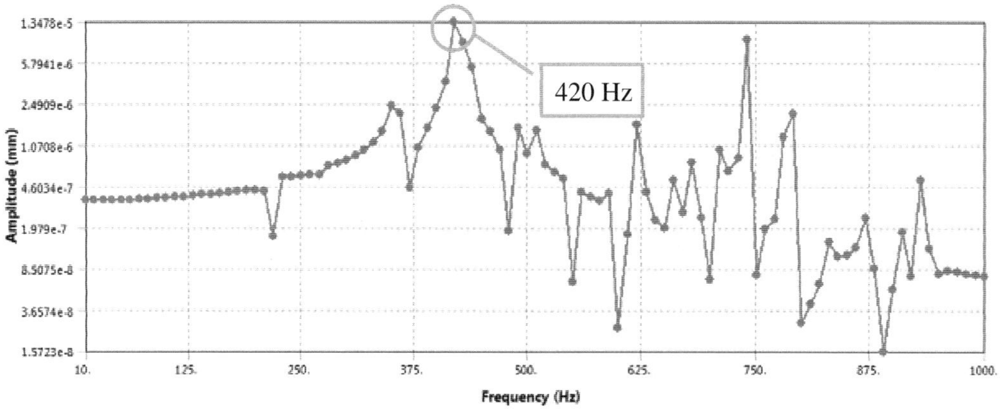
</td></tr>
</table>

3.2. Comparison between MA and ODS of the spillway physical model

Based on the modal analysis result, any vibration or external force frequencies that occurred at 220.87 Hz must be avoided due to the resonance phenomena. Any vibration that occurs at 220.87 Hz frequency will cause the spillway of the dam to undergo failure. Based on figure 9, the ODS result shows that, the deformation amplitude value of 0.003 mm occurred at the operating frequency of 45 Hz and this operating frequency is far from the natural frequency of 220.87 Hz. Thus, in this case the transient vibration will induce minimal effect to the spillway structure and the spillway is considered safe to operate.

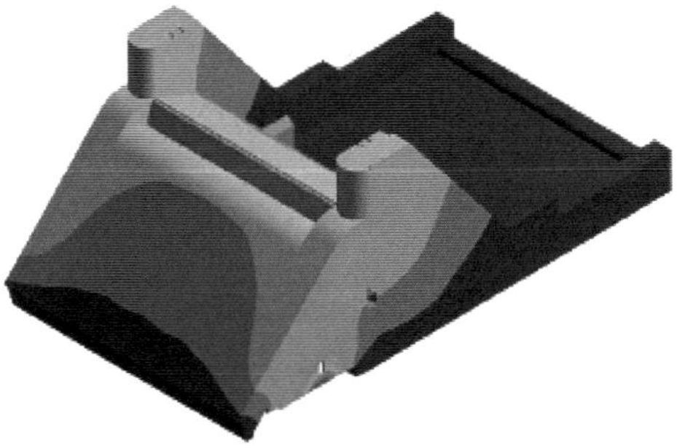

Figure 7. ODS

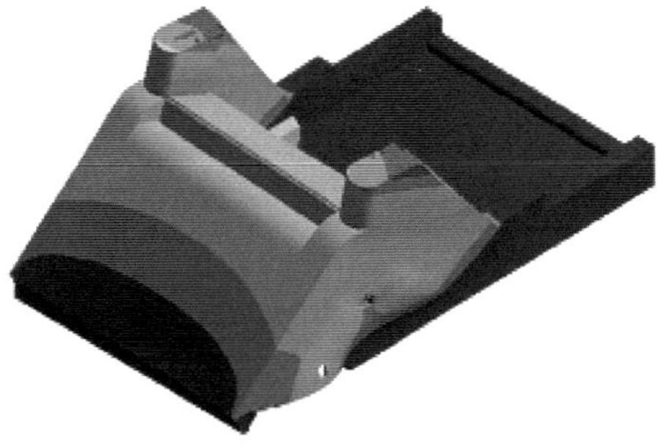

Figure 8. Mode Shape

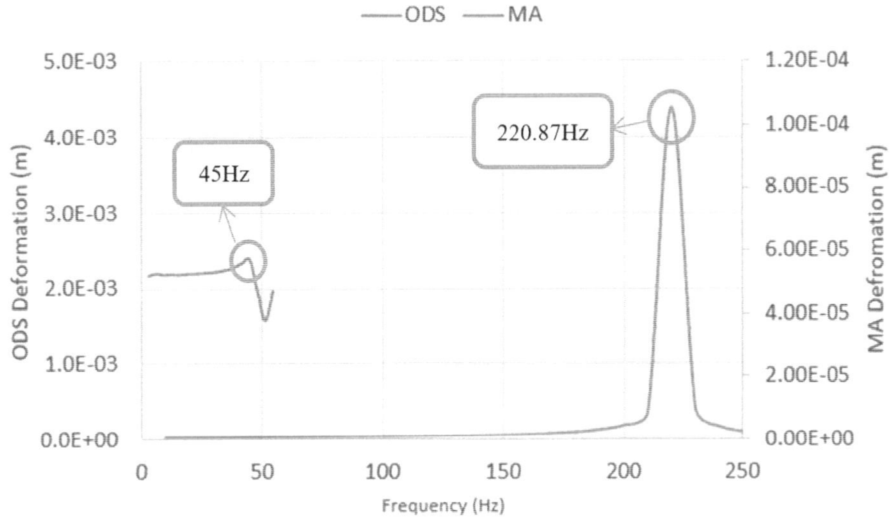

Figure 9. Relationship between ODS and Modal Analysis

3.3. Experimental validation results of spillway physical model

From the EMA conducted at the front divider of the spillway section, the mode shape obtained can be observed in Figure 10 and it occurred at the natural frequency of 273 Hz. Whilst, for ODS, the result of comparison between simulation and experiment is shown in Figure 11. From the result, it can be said that both experiment and simulation are in acceptable range since both are operated in the range of 45 - 52 Hz (13.46 % of error) of operating frequencies and have an almost similar deflection shape. The amplitude for simulation is 0.0024 m whereas for experiment is in the range of 0.001 to 0.0025 m. Therefore, when the water is spilled from the upstream to downstream of the dam, the transient vibration effect at operating frequency of 52 Hz is not coincides with the natural frequency of 273 Hz. This will result in slightly deflection but there is no resonance phenomenon happened.

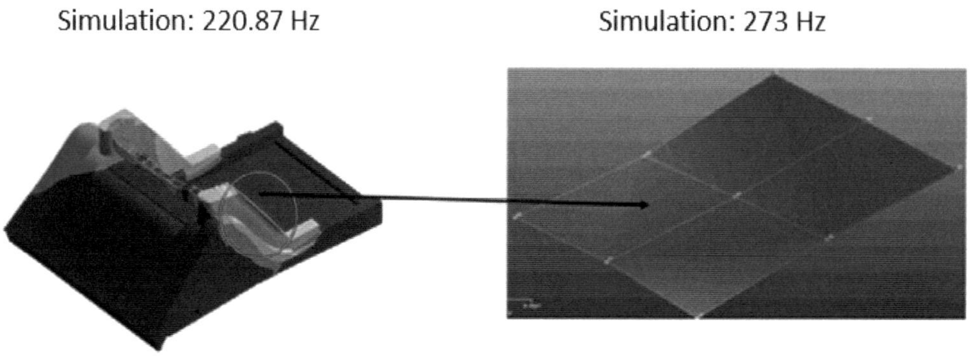

Figure 10. MA of spillway model (front divider) between simulation and experiment

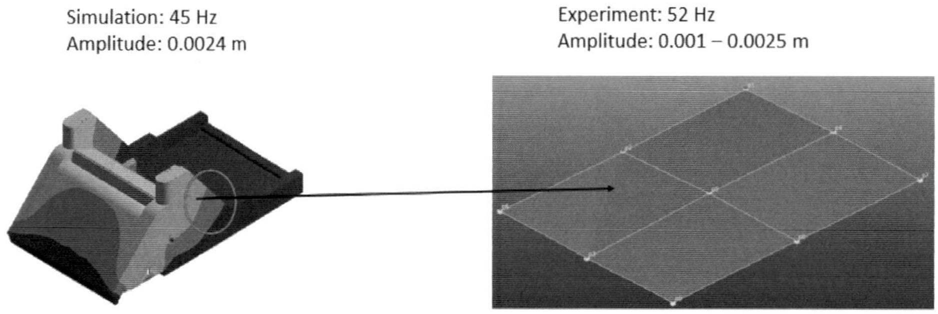

Figure 11. ODS of spillway model (front divider) between simulation and experiment

Table 3 shows the percentage differences between experiment and simulation result for the spillway section (front divider). The percentage differences for MA and ODS is 19.10 % and 13.46 %, respectively, which is generally below an acceptable error percentage of 20 %. From this result, the natural frequencies, mode shapes, operating frequency and ODS for the spillway section shows an agreement with the build physical model.

Table 3. Validation between experimental and simulation results.

	Experiment	Simulation	Percentage Differences
EMA	273 Hz	220.87 Hz	19.10 %
ODS	52 Hz	45 Hz	13.46 %

4. Conclusion

Based on this study, the following conclusions were drawn:

1. The modal analysis of the Chenderoh dam physical model (spillway section) has been successfully conducted. Tetrahedral meshing is used, and six most significant natural frequencies of the spillway are determined. The most dominant natural frequency is at 1st mode of 220.87 Hz with the deflection value 1.6 mm and the lowest natural frequency value is at 424.18 Hz with deflection value of 1.4 mm.
2. The ODS study from FSI of the dam spillway physical model has been successfully carried out. The operating frequency of 45 Hz has been achieved for the ODS with the deformation of 0.0024 m. By comparing the ODS and mode shape, it is realized that, the spillway of the dam is safe to run and there are no resonance phenomena.

3. The experimental validation of MA and ODS have been successfully carried out. The natural frequency and operating frequency obtained from EMA and ODS is 273 Hz and 52 Hz, respectively, which is 19.1 % and 13.46 % of percentage differences compared to the simulation results. The results of both experiment and simulation is acceptable since the percentage differences is below 20 %.

5. References

[1] T. Mazda and Y. Endo, "Static and Dynamic Characteristics of Arch Dams Considering Nonlinear Behavioe of Transverse Joints," *5th World Conf. on. Earthq. Eng.*, 2012.
[2] Y. Zhuan-yun, "Static and Dynamic Analysis of High Arch Dam by Three Dimensional Finite Element Method," *Electron. J. Geotech. Eng.*, vol. 19, pp. 2537–2551, 2014.
[3] M. Mahdizadeh and A. Ghanbari, "Calculation of Natural Frequency of Earth Dams by Means of Analytical Solution," *7th Int. Conf. Case Hist. Geotech. Eng.*, 2013.
[4] A. Debchaudhury, G. Gazetas, and D. A. Gasparini, "Random vibration analysis for the seismic response," *Geotechnique*, no. 2, pp. 261–277, 1981.
[5] W. Shanshan and R. E. N. Qingwen, "Dynamic response of gravity dam model with crack and damage detection," vol. 54, no. 3, pp. 541–542, 2011.
[6] H. Sato and Y. Obuchi, "A study on earthquake responses of actual rock-fill dam and numerical analyses," *12th World Conf. Earthq. Eng.*, 2000.
[7] M. I. Ramli, M. Z. Nuawi, S. Abdullah, M. S. Salleh, U. Teknikal, and H. T. Jaya, "The Study of EMA Effoect on Modal Identification : A Review," vol. 9, no. 1, pp. 103–121, 2017.
[8] S. P. Chaphalkar, S. N. Khetre, and A. M. Meshram, "Modal Analysis of Cantilever Beam Structure Using Finite Element Analysis and Experimental Analysis," *Am. J. Eng. Res.*, vol. 4, no. 10, pp. 178–185, 2015.
[9] M. P. S. V, "Effect of Soil Structure Interaction on Gravity Dam," vol. 4, no. 4, pp. 1046–1053, 2015.
[10] M. Iman and K. A. Abbas, "Dynamic Analysis of Concrete Dam Due to Seismic Forces," vol. 17, no. 8, pp. 1046–1053, 2012.
[11] P. Jia, S. K. Lai, W. Zhang, and C. W. Lim, "Experimental and FEM Modal Analysis of a Deployable-Retractable Wing," *Mod. Mech. Eng.*, no. 4, pp. 183–197, 2014.
[12] J. P. Balsara and J. Fowler, "Vibration tests of North Fork Dam model.pdf." .
[13] Y. Deger, S. Pietrzko, W. Ruecker, and R. G. Rohrmann, "Modal Analysis of a Highway Bridge: Experiment, Finite Element Analysis and Link," pp. 1141–1149.
[14] N. L. Hugar and P. R. Venkatesh, "Shaker based Operating Deflection Shape (ODS) Testing of Two- Wheeler Chassis," *Int. J. Adv. Eng. Res. Dev.*, vol. 1, no. 6, pp. 1–7, 2014.
[15] K. Saravanan and A. Sekhar, "Crack detection in a rotor by operational deflection shape and kurtosis using laser vibrometer measurements," *J. Vib. Control*, vol. 19, no. 8, pp. 1227–1239, 2012.

Acknowledgements

The authors acknowledge UNITEN R & D Sdn Bhd (URND) on the fund provided for this research under the TNB consultancy grant (U-TG-CR-18-04) and Universiti Sains Malaysia through RUI Grant 1001.PEMAKNIK.8014129.

IOP Conf. Series: Materials Science and Engineering **815** (2020) 012021 doi:10.1088/1757-899X/815/1/012021

Fluid/structure Interaction Numerical Study on the Mechanical Integrity of Water Dam Reservoir Banks

F C Ng[1], Aqil Azman[2], Aizat Abas[2*], Mohd Hafiz Zawawi[3], Ahmad Zhafran Ahmad Mazlan[2], Nazirul Mubin Zahari[3] and Z Mohd Remy Rozainy[4]

[1] Institute of Energy Infrastructure (IEI), Universiti Tenaga Nasional, 43000 Kajang, Selangor, Malaysia.
[2] School of Mechanical Engineering, Universiti Sains Malaysia, Engineering Campus, Nibong Tebal, 14300, Penang, Malaysia.
[3] Department of Civil Engineering, College of Engineering, Universiti Tenaga Nasional, 43000 Kajang, Selangor, Malaysia.
[4] School of Civil Engineering, Universiti Sains Malaysia, Engineering Campus, Nibong Tebal, 14300, Penang, Malaysia.

*Corresponding author: aizatabas@usm.my

Abstract. Dam reliability analysis is conducted to access the integrity of dam structure and thus to prevent dam failure. This paper presents the numerical reliability study of reservoir banks of water dam by the mean of fluid/structure interaction (FSI) simulation. Through FSI, the hydrostatics effect of the water in reservoir on the wall of dam was successfully simulated and identified. From the current numerical work, relatively stagnant water on the dam bank was observed, with the increasing water pressure along the water depth. Subsequently, localized high stress and deformation on the dam structure was successfully identified, and suitable countermeasures were being recommended. For the current investigated dam bank structure, the highest deformation and stress detected are respectively 1.185 mm and 3.12 MPa. Therefore, at present operating configuration, it is concluded the bank would not exhibit any failures.

1. Introduction

Dam is the most vital construction infrastructure to human civilization. It reserves water for daily usages and generation of electricity in hydro power plant. However, dam failure, such as dam break and overflow, can causes catastrophic damages, involving hundreds of human lives and their houses apart from economical losses. Therefore, it is common practice that the dam and reservoir are regularly maintained to prevent breakage and able to sustain the hydrostatics pressure [1, 2].

Various literatures had been reported in researching the dam reliability [3 – 11]. Most researchers utilized the numerical methodology to simulate the water flow in the dam and later investigate its structural reliability. Example of numerical scheme used in the dam simulation by past researches are: HEC-ResSim [3], finite element method (FEM) in ABAQUS [4], volume of fluid (VOF) method [5] and fluid/structure interaction (FSI) [6 – 11].

It was found that the FSI approach capable in capturing both hydrodynamic and mechanical parameters simultaneously. Therefore, FSI enables a more actual dam phenomenon being modelled numerically. FSI can be easily executed using the commercially available finite volume method (FVM) software, ANSYS, using the packages of Fluent, Mechanical and System Coupling [12, 13]. Previous dam's radial gates FSI simulations used the commercially available ANSYS software [6, 7, 8, 11]. Other than ANSYS, FSI simulation can also be implemented using in-house finite element code, SNACS [10].

In this paper, FSI study is conducted on the reservoir bank of a dam, to investigate the reliability of the concrete dam walls upon being exerted by the hydrostatics pressure. Various data (e.g. velocity, pressure, deformation and stress) will be analyzed to identify the reliability of dam and whether if any failure shall occur. The primary goal of current simulation work is to present the modelling of FSI phenomenon dam wall, which may be useful in predicting potential dam failure.

2. Governing equations

2.1 Fluid phase

In the current simulation of water flow in dam reservoir banks, the three-dimensional, incompressible Navier-Stokes equations were adopted with pressure-based solver. Incompressible Navier-Stokes equations decouple the energy equation by constraining the density to be constant, which is a reasonable assumption for the water. The emphasis of the simulation is on the pressures exerted on the dam structure and not the heat transfer. Hence, the energy equation could be decoupled from the mass and momentum equations without compromising the accuracy of the simulation results. With the energy equation discarded, the compact governing continuity and momentum equations can be respectively written as follows:

$$\frac{\partial \rho}{\partial t} + \nabla \cdot (\rho \vec{v}) = S_m \, , \tag{1}$$

$$\frac{\partial}{\partial t}(\rho \vec{v}) + \nabla \cdot (\rho \vec{v} \cdot \vec{v}) = -\nabla p + \nabla \cdot \bar{\bar{\tau}} + \rho \vec{g} + \vec{F}. \tag{2}$$

where the source term, S_m is the mass added to the continuous phase from the dispersed second phase and any user-defined source, which is zero in the simulations; p is the static pressure; the term of $\rho \vec{g}$ denotes the gravitational body force and $\vec{F}$ is the external body force (i.e. that arise from interaction with the dispersed phase). The stress tensor $\bar{\bar{\tau}}$ is given by:

$$\bar{\bar{\tau}} = \mu \left[(\nabla \vec{v} + \nabla \vec{v}^T) - \frac{2}{3} \nabla \cdot \vec{v} I \right], \tag{3}$$

where μ is the molecular viscosity, I is the unit tensor, and the term on the right-hand side is the effect of volume dilation.

The turbulence model employed was the standard k-ε model [12]. The standard k-ε turbulence model falls within the class of two-equation turbulence models which allow the determination of both, a turbulent length and time scale by solving two separate transport equations, as follows:

$$\frac{\partial}{\partial t}(\rho k) + \frac{\partial}{\partial x_i}(\rho k u_i) = \frac{\partial}{\partial x_j}\left[\left(\mu + \frac{\mu_t}{\sigma_k}\right)\frac{\partial k}{\partial x_j}\right] + G_k + G_b - \rho\varepsilon - Y_M + S_k, \tag{4}$$

$$\frac{\partial}{\partial t}(\rho \varepsilon) + \frac{\partial}{\partial x_i}(\rho \varepsilon u_i) = \frac{\partial}{\partial x_j}\left[\left(\mu + \frac{\mu_t}{\sigma_\varepsilon}\right)\frac{\partial \varepsilon}{\partial x_j}\right] + C_{1\varepsilon}\frac{\varepsilon}{k}(G_k + G_{3\varepsilon}G_b) - C_{2\varepsilon}\rho\frac{\varepsilon^2}{k} + S_\varepsilon. \tag{5}$$

In equations (4) and (5), G_k represents the generation of turbulence kinetic energy due to the mean velocity gradients, G_b is the generation of turbulence kinetic energy due to buoyancy and Y_M represents the contribution of the fluctuating dilatation in compressible turbulence to the overall dissipation rate. Additionally, σ_k and σ_ε are respectively the turbulent Prandtl's numbers for k and ε; S_k and S_ε are user-defined source terms and lastly with constant terms of $C_{1\varepsilon}$, $C_{2\varepsilon}$ and $C_{3\varepsilon}$. Detailed interpretations of these terms and their formulation are provided in the ANSYS Fluent documentation [12].

Subsequently, to account for the air-water interactions in the dam simulation, the multiphase Volume of Fluid (VOF) model was employed. VOF can model two or more immiscible fluids by solving a single set of momentum equations and tracking the volume fraction of each of the fluids throughout the domain. The volume fraction equation is given as:

$$\frac{1}{\rho_q}\left[\frac{\partial}{\partial t}(\alpha_q\rho_q) + \nabla\cdot(\alpha_q\rho_q\vec{v}_q)\right] = \frac{S_{\alpha_q}}{\rho_q} + \frac{1}{\rho_q}\sum_{p=1}^{n}(\dot{m}_{pq} - \dot{m}_{qp}), \tag{6}$$

where $\dot{m}_{qp}$ is the mass transfer from phase q to phase p whereas $\dot{m}_{pq}$ is the mass transfer from phase p to phase q.

2.2 Structural phase

The structure simulations were carried out with ANSYS Mechanical [13]. After the pressure loads were acquired from the fluid simulations, the results were fed as input to static structural analysis for the dam gates. For static linear analyses, ANSYS Mechanical APDL applied the following equation:

$$[K]\{u\} = \{F^a\} + \{F^r\}, \tag{7}$$

where $\{u\}$ is the nodal degree of freedom (DOF) vector, $[K]$ is the total stiffness or conductivity matrix, which defined as

$$[K] = \sum_{m=1}^{N}[K_e], \tag{8}$$

with number of elements, N and element stiffness or conductivity matrix, $[K_e]$. On the right hand side of equation (7), $\{F^r\}$ is the nodal reaction load vector and $\{F^a\}$ is the total applied load vector, which is the sum of applied nodal load vector, $\{F^{nd}\}$ and total of all element load vector effects (pressure, acceleration, thermal, gravity), $\{F^e\}$:

$$\{F^a\} = \{F^{nd}\} + \{F^e\}. \tag{9}$$

Finite element (FE) procedures are applied on all structural models by ANSYS Mechanical to generate system of simultaneous linear equations which have been described in the previous section. These equations are then solved by either a direct elimination process or an iterative method. As in current dam simulation, the Preconditioned Conjugate Gradient (PCG) solver was chosen.

PCG solver is one of the three iterative methods available in ANSYS Mechanical APDL which is efficient and reliable for all types of analyses including the ill-conditioned beam/shell structural analysis. The PCG solver is only valid for real symmetric stiffness matrices. To solve typical system of equations in the form of previous equation iteratively, general CG methods cast the solution in the form of a series of vectors $\{p_i\}$:

$$\{u\} = \alpha_1\{p_1\} + \alpha_2\{p_2\} + \alpha_3\{p_3\} + \cdots + \alpha_m\{p_m\}. \tag{10}$$

3. Numerical simulations

The fluid/structure interaction (FSI) of water flow and dam's reservoir bank (as in Figure 1) is numerically simulated using the commercially available software, ANSYS. The three-dimensional flow is regarded as incompressible, unsteady and turbulence. In FLUENT settings, k-ε turbulence model and multiphase VOF model were applied together with the pressure-based solver. Besides, the implicit time scheme is employed, with fixed time step of 0.01 s being chosen for the transient flow simulation. In the

System Coupling, data transfers were created for all the fluid-structure interaction regions for both fluid and structural domains.

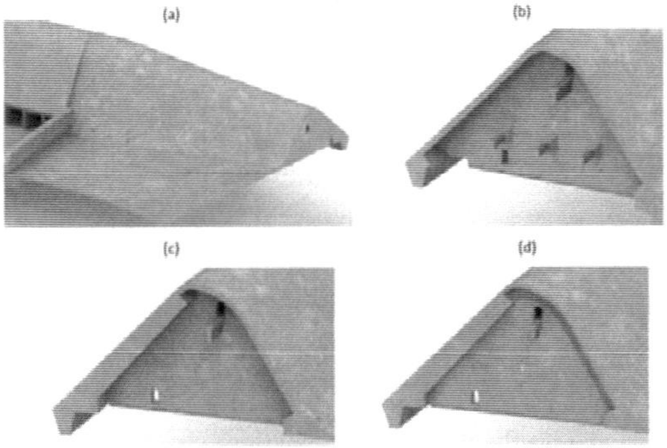

Figure 1. Geometrical details of the dam's reservoir bank.

For the aspect of boundary conditions (BCs) considered, the flow inlet is set as the velocity inlet of 1.0 m/s; while the flow outlet is designated as the pressure outlet at ambient atmospheric pressure of 1 atm. Subsequently, the fluid/structure surfaces are set as no-slip walls. Gravity is being considered with magnitude of 9.81 m/s^2 to mimic the actual application and thus account for the hydrostatics effect. On other hand, for the BC in structural domain, the base of reservoir bank is set to be fixed support such that it would remain stationary throughout the flow; and fluid-structure interaction (FSI) for the solid structure surface that in contact with the fluid phases. The phases that exist in the fluid domain are air and water, with both at temperature of 25°C and pressure of 1 atm. In multiphase VOF scheme, the primary phase is air and secondary phase is incompressible water. The dam is made of concrete material with Young modulus of 5.53 GPa, density of 2300 kg/m^3, tensile ultimate strength of 5 MPa and Poisson's ratio of 0.18. While the material properties of air and water are specified in the material data library of ANSYS. Due to the complexity of the geometry, hybrid unstructured meshes (consists of the hexagonal and tetrahedral meshes) were generated for both the fluid and structural domains (refers Figure 2). The mesh resolution is set at fine, to balance off both the accuracy and computational time.

Figure 2. Generated hybrid mesh for the structural reservoir bank.

4. Results and Discussion

The fluid/structure interaction (FSI) simulation on the reservoir bank had yielded various hydrodynamics data (velocity and pressure) and mechanical data (deformation and stress). In Figure 2, the velocity of water in the reservoir is found to be within the range of 0 m/s (completely stagnant water) to 4.5 m/s. Such nearly stagnant flow is due absent of overtopping flow and the water remains in the reservoir. These stagnant waters stored in the reservoir would be exerted hydrostatics pressure to the wall of the dam, which as shown in the Figure 3. The pressure gradually increases as it gradually descends to the base of the dam, as depicted in Figure 4.

On the structural aspect, the deformation contour in Figure 5 shows that there is a localized high deformation region with value of 1.185 mm. Such high deformation is occurred at the dam section divider. Therefore, regular maintenance and checking must be conducted particularly on this region, for any sign of crack propagations. Besides, the stress contours on Figure 6 revealed that the walkway linking the reservoir bank and bottom outlet exhibited highest stress. Furthermore, from the cross-sectional cut in Figure 6, it is found that the interior part of the dam has slightly higher stress compared to that at the dam outer surfaces. However, the maximal stress of the whole reservoir dam is identified as 3.12 MPa, which is substantially lower than the concrete yield stress of 20 MPa. Therefore, the reservoir bank of the dam would not fail, at current present state of water level and flow.

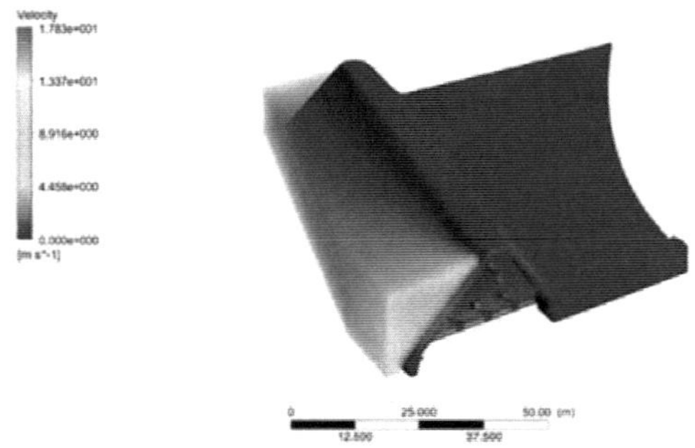

Figure 3. Velocity contour of the water flow near the reservoir bank.

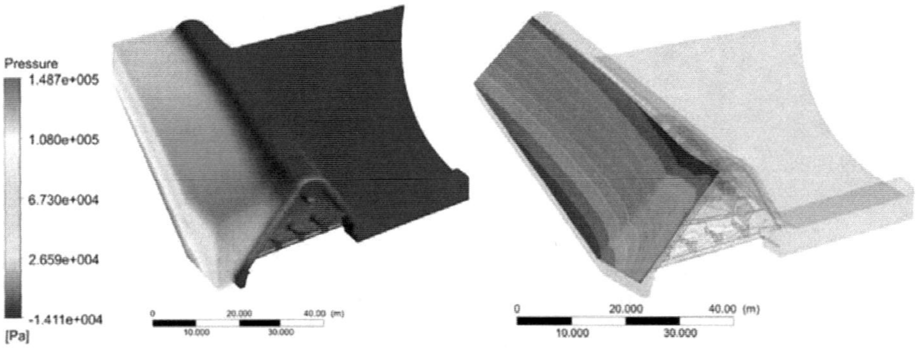

Figure 4. Gauge pressure contours of the water flow near the reservoir banks. The right contour depicts the pressure distribution of water on the dam wall.

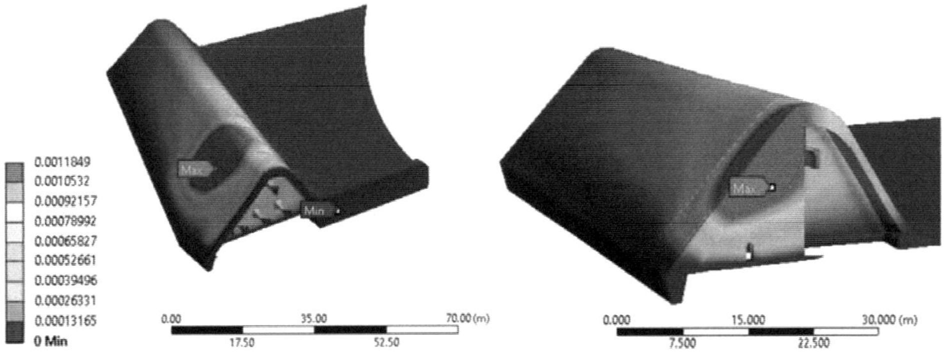

Figure 5. Deformation contours (in unit meter) of the reservoir bank's concrete wall. The right contour depicts the cross-sectional cut of dam wall where high deformation being recorded.

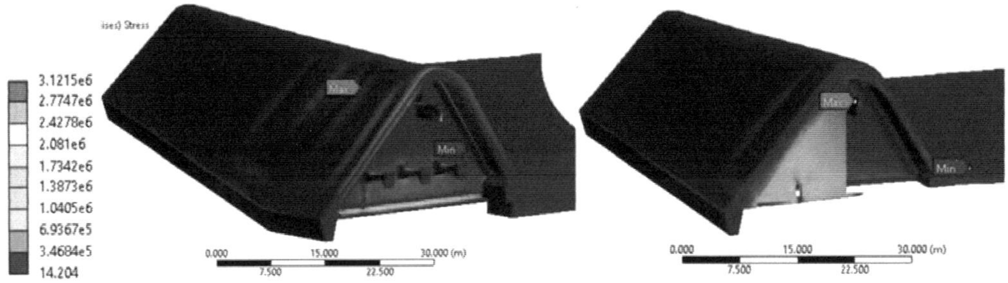

Figure 6. Stress contours (in unit Pa) of the reservoir bank's concrete wall. The right contour depicts the cross-sectional cut of the dam wall, to visualize the interior stress.

5. Conclusions

By means of fluid/structure interaction (FSI), that utilized both multiphase VOF and k-ε models, both the hydrodynamic and mechanical aspects of reservoir bank of a dam had been successfully investigated. The water profile is found to remain relatively stagnant due to the absent of overtopping. Besides, the water pressure exerted on the wall of dam is gradually increasing toward the base of the reservoir. A high deformation of 1.18 mm is noticed on the section divider of the dam. However, the overall stress level of the dam upon exerted by the water reservoir is well below the yield stress of concrete, such that it would not fail. Generally, through the detailed FSI numerical approach, the mechanical reliability aspect of the reservoir bank of a dam can be studied alongside with the fluid flow of water in the dam. Moreover, the FSI numerical simulation was demonstrated viable for the dam reliability study.

6. References

[1] L M Sidek, F C Ros and N H A. Aziz 2011 *Student Conference on Research and Development*
[2] S Solava and N Delatte 2003 *Proceedings of the Third ASCE Forensics Congress*
[3] G Uysala, B Akkol, M. Irem Topcu, A. Sensoy and D. Schwanenberg 2016 *Procedia Engineering* **154** 1385–92.
[4] Q Zhang, D Li, F Wang and B Li 2018 *Engineering Failure Analysis* **91** 72–91
[5] A. Issakhov, Y. Zhandaulet and A Nogaeva 2018 *International Journal of Multiphase Flow*
[6] F C Ng, A Abas, I Abustan, Z M R Rozainy, M. Z. Abdullah, A. B. Jamaludin and S. M. Kon 2018 *IOP Conf. Ser. Mater. Sci. Eng.* **370**(1) 012062
[7] F C Ng, A Abas, I Abustan, Z M R Rozainy, M. Z. Abdullah, A. B. Jamaludin, and S. M. Kon 2018 *IOP Conf. Ser. Mater. Sci. Eng.* **370**(1) 012063

[8] M N Nashurdin, A Abas, A Azman, F C Ng, M R M Radzi and A Hassani 2019 *AIP Conference Proceedings* **2129** 020044
[9] N. Cheraghi-Shirazi, A.R. Kabiri-Samani and B. Boroomand 2014 *Flow Measurement and Instrumentation* **40** 91–8
[10] O Omidi and V Lotfi 2017 *Journal of Fluids and Structures* **69** 34–55
[11] T B Amina, B Mohamed, L André and B Abdelmalek 2015 *Engineering Structures* **84** 19–28
[12] ANSYS Fluent Documentation 2017 *ANSYS Inc.*
[13] ANSYS Mechanical Documentation 2017 *ANSYS Inc.*

Acknowledgments
This research work is supported by the Fundamental Research Grant Scheme (FRGS) (Grant number: FRGS/1/2019/TK03/USM/03/1).

IOP Conf. Series: Materials Science and Engineering **815** (2020) 012022 doi:10.1088/1757-899X/815/1/012022

Coastal Sediment Transport Study along Shoreline by using Smooth Particle Hydrodynamics

M H Zawawi[1*], Aqil Azman[3], M R M Radzi[2], Aizat Abas[3] and Nurhanani A Aziz[1]

[1]Uniten R&D Sdn Bhd (URND), Universiti Tenaga Nasional, Selangor, Malaysia
[2]TNB Generation Division, Tenaga Nasional Berhad, 59200 Kuala Lumpur, Malaysia
[3]School of Mechanical Engineering, Universiti Sains Malaysia, Engineering Campus, Nibong Tebal, 14300, Penang.

*Corresponding author: MHafiz@uniten.edu.my

Abstract. Sediment transport typically due to combination of gravity acting on the sediment, and/or the movement of the fluid in which the sediment is entrained. This paper gives an exposure simulation of three-dimensional (3D) sediment transports by using Smooth Particle Hydrodynamics (SPH) at coastal area. The SPH formulation from Monaghan constructed in the code is modified according to the desired model to be solved. The code is then computed by using single Graphics Processing Unit (GPU) with 14 cores multiprocessors. The behavior of sediment transport is investigated in this research based on Herschel-Bulkley-Papanastasiou (HBP) model. A constant sinewave of $0.5\sin(0.5\pi t)$ is applied to the fluid domain to create sea wave effect that will cause sediment transport. The results show that the goal to get sediment transport is achieved and for the further study, the rate of erosion and accretion can be determined to give more meaningful and valuable to the result.

1. Introduction

Sediment transport is the one of the problems occur in coastal area which is can cause catastrophic failure to the building that build near to the coastal area. Some country builds power plants near the coastal area to extract natural resources such as natural gas. The study of sediment transport is important in order to make sure that the building is safe for several or long periods of time. The simulation of sediment transports are difficult to model by Eulerian approach (traditional mesh) as compared to Lagrangian based approach [1]. SPH was originally developed to study astrophysical problems [2, 3, 4] and has been widely used now a day to the engineering applications. Previously, the study of sediment transport by using SPH is more focused on two-dimensional (2D) as compared to 3D. Manenti et al. (2012) used 2D SPH simulation to model sediment flushing induced by a rapid water flow [5,6].

Besides that, Shi et al. (2017) also used 2D SPH model for sediment laden flows [7]. After that, the 2D multi-phase SPH model is improved to investigate scouring and erosion [8, 9]. In addition, Ghaintanellis et al. (2018) used an elastic-viscoplastic model [10] is combined to a multi-phase SPH formulation [11, 12] to model granular flows and non-cohesive sediment transport [13]. Furthermore, the 3D model of SPH is effectively used in numerical analysis study of Sarawak barrage river bed erosion and scouring [14-17]. Therefore, this paper attempts to avoid such boundaries with the aim of producing high-precision 3D simulation model of sea wave-sediment interaction with minimum computational effort.

2. SPH formulation

The SPH involves the computation of discretized fluid into a set of particles, each particle is referred to a nodal point in which physical quantities are computed as an interpolation of the values of the nearest particles. The integral approximation of any function A($\vec{r}$) is

$$A(\vec{r}) = \int_{\Omega} A\left(\vec{r'}\right) W\left(\vec{r} - \vec{r'}, h\right) dr' \quad (1)$$

In which h is the smoothing length and $W\left(\vec{r} - \vec{r'}, h\right)$ is the smoothing kernel or weighting function at particle a is given by:

$$A(\vec{r}) = \sum_b m_b \frac{A_b}{\rho_b} W_{ab} \quad (2)$$

In which m_b and ρ_b being the mass and the density associated to the neighboring particle b, respectively. The term W_{ab} describes the weight function which is $W_{ab} = W(\vec{r_a} - \vec{r_b}, h)$. There are several types of kernel definitions and for this research, the cubic spline kernel is selected and shown as follows:

$$W(r,h) = \alpha_d \begin{cases} 1 - \frac{3}{2}q^2 + \frac{3}{4}q^3 & 0 \le q \le 1 \\ \frac{1}{4}(2-q)^3 & 1 \le q \le 2 \\ 0 & q \ge 2 \end{cases} \quad (3)$$

In which $\alpha_d = \frac{1}{\pi h^3}$ in three-dimensional.

The momentum equation in SPH method is implemented to determine the acceleration of a particle a as the result of the particle interaction with particle b:

$$\frac{d\vec{v_a}}{dt} = -\sum_b m_b \left(\frac{P_b}{\rho_b^2} + \frac{P_a}{\rho_a^2} \Pi_{ab}\right) \nabla_a W_{ab} + \vec{g} \quad (4)$$

The detailed SPH formulation can be obtained in book [18].

3. Methodology

3.1. Modelling of Coastal Area

3.1.1. Coastal modelling by using Solidworks

The model is constructed in Solidworks based on the scanned topography from the drone as shown in Fig. 1.

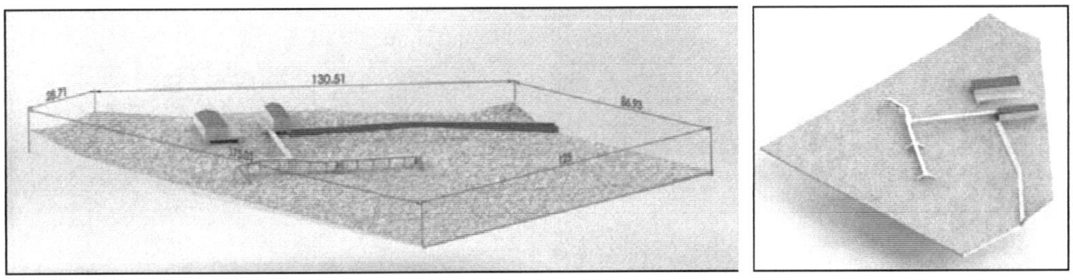

Figure 1. The constructed model by using Solidworks.

3.1.2. SPH model

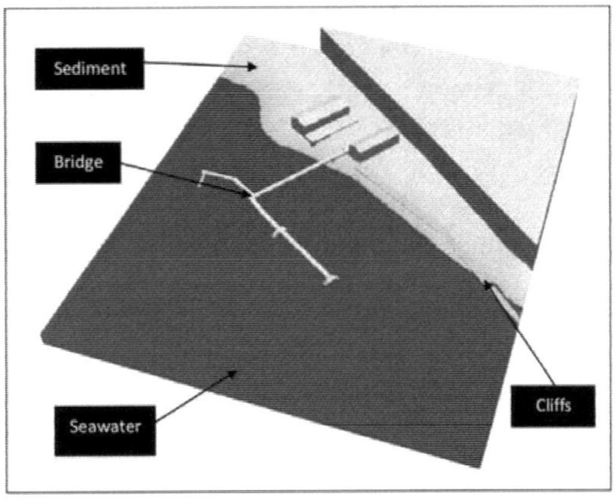

Figure 2. SPH model for the coastal area.

Fig 2 shown a SPH model for the numerical analysis whereby elements such as sediment, bridge, seawater and cliff are represented. The wave is generated by the wave paddle is 0.5sin(0.5πt) to mimic the real case study.

4. Results and Discussions

Fig 3 shown that the changes of coastal profile along the duration of 100s. The red circles are the area that can be noticed the movement of the sediment. The movement of sediment highly occurs at the region near to the building due to the wave erosion.

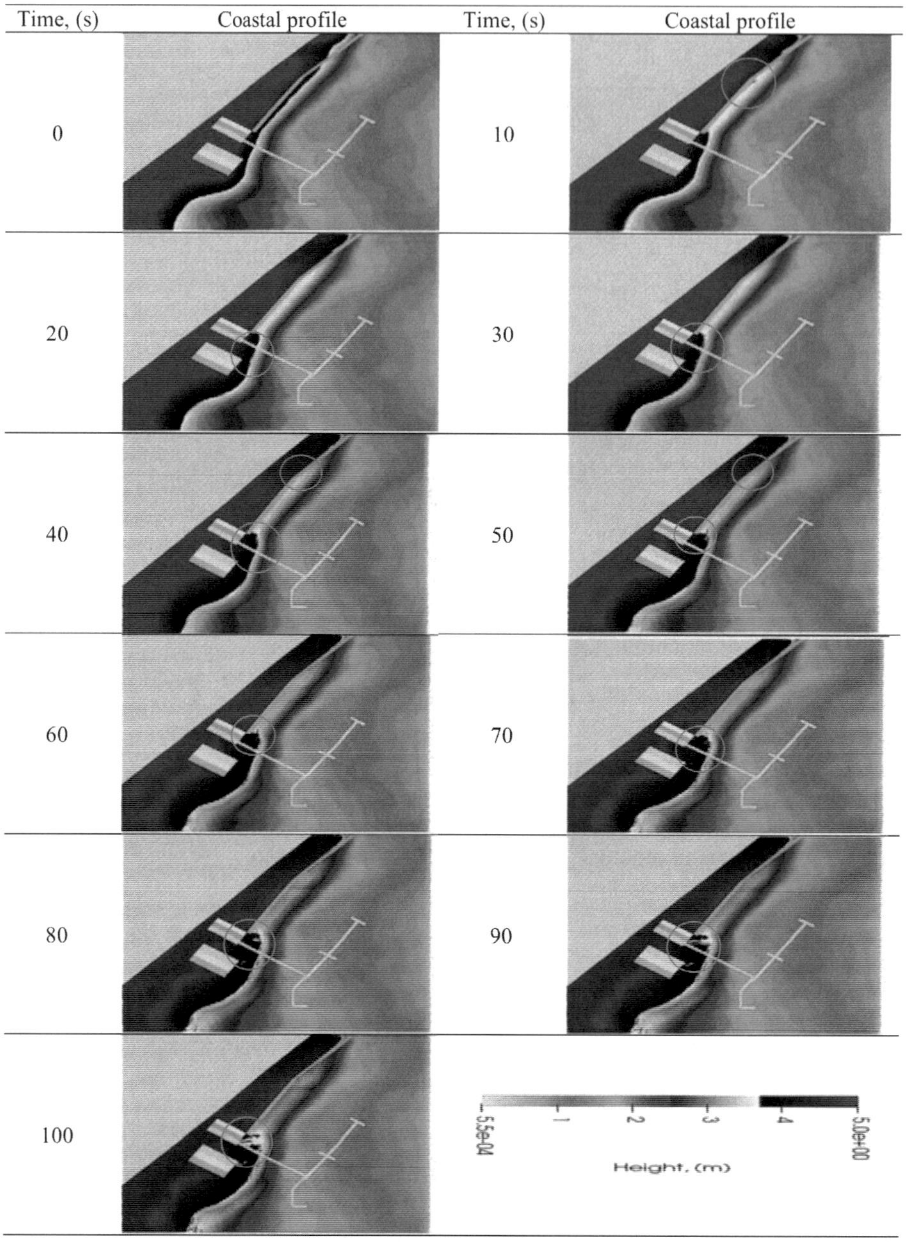

Figure 3. Coastal profile changes within the 100s periods.

5. Conclusions

In this paper, 3D simulation of SPH is combined with the wave generation equation to simulate liquid-sediment for application to coastal area. This 3D simulation was shown effectively simulate the effect of sediment transport due to interaction of sea wave. This allows early precaution to be taken to avoid any failure to the building near to the coastal area. In the near future, the rate of erosion and accretion can be determined to give more valuable and meaningful data to the result.

IOP Conf. Series: Materials Science and Engineering **815** (2020) 012022 doi:10.1088/1757-899X/815/1/012022

6. References

[1] Müller M, Charypar D and Gross M 2003 Particle-Based Fluid Simulation for Interactive Applications *Proceedings of the 2003 ACM SIGGRAPH/Eurographics symposium on Computer animation* pp 154–9

[2] Gingold R A and Monaghan J J 1977 Smoothed particle hydrodynamics: theory and application to non-spherical stars *Mon. Not. R. Astron. Soc.* **181** 375–89

[3] Lucy L B 1977 A Numerical Approach to The Testing of The Fission Hypothesis *Astron. J.* **82** 1013–24

[4] M. Radzi, Nashruddin, M.N, Zawawi, M. H., Abas, Aizat, Azman, Aqil and Hassani, A, "Effect of free surface water level of Kenyir Dam on spillway structure using fluid-structure interaction (FSI) analysis," Applied Physics of Condensed Matter (Apcom 2019), vol. 2131, no. August, 2019.

[5] Manenti S, Sibilla S, Gallati M, Agate G and Guandalini R 2012 SPH Simulation of Sediment Flushing Induced by a Rapid Water Flow *J. Hydraul. Eng.* **138** 272–84

[6] N. M. Zahari, Zawawi, M. H., Sidek, L. M., Mohamad, Daud, Itam, Zarina, Ramli, M. Z., Syamsir, Agusril, Abas, Aizat and Rashid, M., "Introduction of discrete phase model (DPM) in fluid flow: A review," in AIP Conference Proceedings 2030, 2018.

[7] Shi H, Yu X and Dalrymple R A 2017 Development of a two-phase SPH model for sediment laden flows *Comput. Phys. Commun.* **221** 259–72

[8] Fourtakas G and Rogers B D 2016 Modelling multi-phase liquid-sediment scour and resuspension induced by rapid flows using Smoothed Particle Hydrodynamics (SPH) accelerated with a Graphics Processing Unit (GPU) *Adv. Water Resour.* **92** 186–99

[9] Zubeldia E H, Fourtakas G, Rogers B D and Farias M M 2018 Multi-phase SPH model for simulation of erosion and scouring by means of the shields and Drucker–Prager criteria. *Adv. Water Resour.* **117** 98–114

[10] Ulrich C, Leonardi M and Rung T 2013 Multi-physics SPH simulation of complex marine-engineering hydrodynamic problems *Ocean Eng.* **64** 109–21

[11] Hu X Y and Adams N A 2006 A multi-phase SPH method for macroscopic and mesoscopic flows *J. Comput. Phys.* **213** 844–61

[12] Ghaitanellis A, Violeau D, Leroy A, Joly A and Ferrand M 2015 Application of the unified semi-analytical wall boundary conditions to multi-phase SPH *10th International SPHERIC Workshop, Parma, Italy*

[13] Ghaïtanellis A, Violeau D, Ferrand M, Abderrezzak K E K, Leroy A and Joly A 2018 A SPH elastic-viscoplastic model for granular flows and bed-load transport *Adv. Water Resour.* **111** 156–73

[14] Zainol M R R M A, Kamaruddin M A, Zawawi M H and Wahab K A 2017 Numerical analysis study of sarawak barrage river bed erosion and scouring by using Smooth Particle Hydrodynamic (SPH) *IOP Conference Series: Materials Science and Engineering* vol 267

[15] M. H. Zawawi, Saleha, A., Salwa, A., Hassan, N. H., Zahari, N. M., Ramli, M. Z. and Muda, Z. C., "A review: Fundamentals of computational fluid dynamics (CFD)," in Green Design and Manufacture: Advanced and Emerging Applications, 2018.

[16] A. Azman, Abas, Aizat, Zawawi, M. H., Rozainy M. A. Z., Mohd. Remy and Abustan, Ismail, "Numerical study of the effect of flow on bottom outlet of dam," Applied Physics of Condensed Matter (Apcom 2019), vol. 2131, no. July, p. 020090, 2019.

[17] M. H. Zawawi, Hazwan, Mohamad, Ghazali, Mohd, Zhafran, Ahmad and Mazlan, Ahmad, "Vibration analysis due to frequent spilling over hollow buttress Chenderoh Dam sector gate spillway," in 2018 International Conference on Vibration, Sound and System Dynamics (ICVSSD 2018), 2019.

[18] Violeau D 2012 *Fluid mechanics and the SPH method: Theory and applications* (Oxford University Press)

IOP Conf. Series: Materials Science and Engineering **815** (2020) 012022 doi:10.1088/1757-899X/815/1/012022

Acknowledgements
The authors acknowledge UNITEN R & D Sdn Bhd (URND) on the fund provided for this research under the TNB consultancy grant (U-TG-CR-18-04).

IOP Conf. Series: Materials Science and Engineering **815** (2020) 012023 doi:10.1088/1757-899X/815/1/012023

Smooth Particle Hydrodynamics Study of Sediment Profile Erosion

Aqil Azman[1], M H Zawawi[2*], M R M Radzi[3], Aizat Abas[4] and Nurhanani A Aziz[2]

[1]Uniten R&D Sdn Bhd, Universiti Tenaga Nasional (UNITEN), Kajang 43000, Selangor Darul Ehsan, Malaysia.
[2]Department of Civil Engineering, College of Engineering, Universiti Tenaga Nasional (UNITEN), Kajang 43000, Selangor Darul Ehsan, Malaysia.
[3]Tenaga Nasional Berhad Generation, Asset Management Department, No. 129, Jalan Bansar, 59200 Kuala Lumpur, Malaysia.
[4]School of Mechanical Engineering, Universiti Sains Malaysia, Engineering Campus, Nibong Tebal, 14300, Penang.

*Corresponding Author: MHafiz@uniten.edu.my

Abstract. A sediment profile erosion with interaction of water which moving with wave generator are model in three-dimensional (3D) by using numerical method Smooth Particle Hydrodynamics (SPH). The code is constructed using SPH formulation from Monaghan and both sediment and water are modelled as weakly compressible viscous fluids. The wave generator is set with a constant sinewave of $0.24\sin(2.6\pi t)$. The sediment model with slope angle 36.464° is constructed in Solidworks with boundary length of 1206 mm, width of 454 mm and height of 463 mm. Meanwhile, the model is simulated in period of 15s. A sensitivity analysis is performed in order to add degree of trustiness to the results. The impact of generated wave to the sediment profile is analysed. The result show that the sediment starts eroded at 3 s. Afterwards, the sediment continued eroded whereby the sediment starts to move from the top region to lower region. This phenomenon due to swash and backwash of the wave that being generated by wave generator. Thus, the results show that the sediment erosion successfully simulated by using SPH.

1. Introduction

Now a day, sediment erosion is the one of the phenomena which is the most concern by researcher. The classical study of sediment erosions used experiment to model the sediment erosion which are difficult to analyse precisely the data as compared to simulation. There are several study using simulation to model sediment erosion shows the capability of Lagrangian based approach as compared to Eulerian approach which difficult to model the sediment erosion[1]. Before this, SPH was used to study astrophysical problems**[2, 3]**. Now a day many researchers start to use simulation to solve engineering problems. Previously, the study of sediment erosion by using SPH is more focused on two-dimensional (2D). As example, Manenti et al. (2012) used 2D SPH simulation in their study to model sediment flushing induced by a rapid water flow[4]. After that, Shi et al. (2017) also used 2D SPH model for sediment laden flows[5]. Furthermore, Fourtakas and Rogers (2016) and Zulbedia et al. (2018) improved the 2D multi-phase SPH model to investigate scouring and erosion**[6, 7]**. Next, Ghaintanellis et al. (2018) used a multi-phase SPH formulation**[9, 10]** combined with an elastic-viscoplastic model[8] to model non-cohesive sediment transport and granular flows[11]. In addition, Zainol et al. (2017) model a 3D SPH in numerical analysis study of Sarawak barrage river bed erosion and scouring effectively[12]. Therefore, this paper aims of producing high-precision 3D simulation model of sediment erosion with minimum computational effort.

2. SPH formulation

The computation of SPH discretized fluid into a set of particles, each particle is referred to a nodal point in which physical quantities are calculated as an interpolation of the values of the nearest particles. The integral approximation of any function A($\vec{r}$) is

$$A(\vec{r}) = \int_{\Omega} A\left(\vec{r'}\right) W\left(\vec{r} - \vec{r'}, h\right) dr' \qquad (1)$$

In which h is the smoothing length and $W\left(\vec{r} - \vec{r'}, h\right)$ is the smoothing kernel or weighting function at particle a:

$$A(\vec{r}) = \sum_b m_b \frac{A_b}{\rho_b} W_{ab} \qquad (2)$$

In which m_b and ρ_b being the mass and the density associated to the neighboring particle b, respectively. The term $W_{ab} = W(\vec{r_a} - \vec{r_b}, h)$ describes the weight function. There are several types of kernel definitions and for this research, the cubic spline kernel is selected and showed as follows:

$$W(r,h) = \alpha_d \begin{cases} 1 - \frac{3}{2}q^2 + \frac{3}{4}q^3 & 0 \le q \le 1 \\ \frac{1}{4}(2-q)^3 & 1 \le q \le 2 \\ 0 & q \ge 2 \end{cases} \qquad (3)$$

In which $\alpha_d = \frac{1}{\pi h^3}$ in three-dimensional.

The momentum equation in SPH method is implemented to determine the acceleration of a particle a as the result of the particle interaction with particle b:

$$\frac{d\vec{v_a}}{dt} = -\sum_b m_b \left(\frac{P_b}{\rho_b^2} + \frac{P_a}{\rho_a^2} \Pi_{ab}\right) \nabla_a W_{ab} + \vec{g} \qquad (4)$$

More detailed on SPH formulation can be found in book [13].

3. Methodology

3.1. Modelling of sediment

3.1.1. Modelling by using Solidworks

The model is constructed in Solidworks based as shown in Fig. 1., then the constructed model was import to SPH model

Isometric view

Side view

Figure 1. The model constructed by using Solidworks.

3.1.2. SPH model

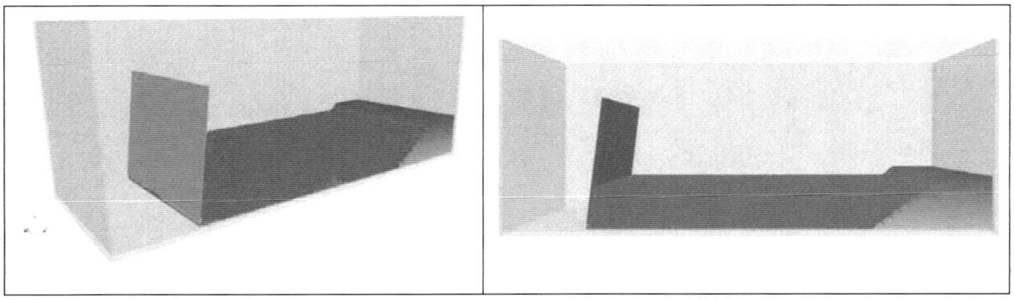

Isometric view Side view

Figure 2. SPH model.

Fig 2 shown a SPH model for the numerical analysis whereby elements such as sediment, wave paddle, water and boundary are represented. The wave is generated by the wave paddle is $0.24\sin(2.6\pi t)$ to mimic the real case study.

4. Results and Discussions

Fig 3 shown that the sediment profile changes along the duration of 15s. During the period of 3s, the sediment starts to erode and moving out to boundary of the initial sediment profile line. After that, the top of sediment moving downward due to backwash from the wave at 6 s. Next, the sediment continuously erodes from 9 s to 15 s and the top of the sediment also keep moving downward due to the backwash from the wave.

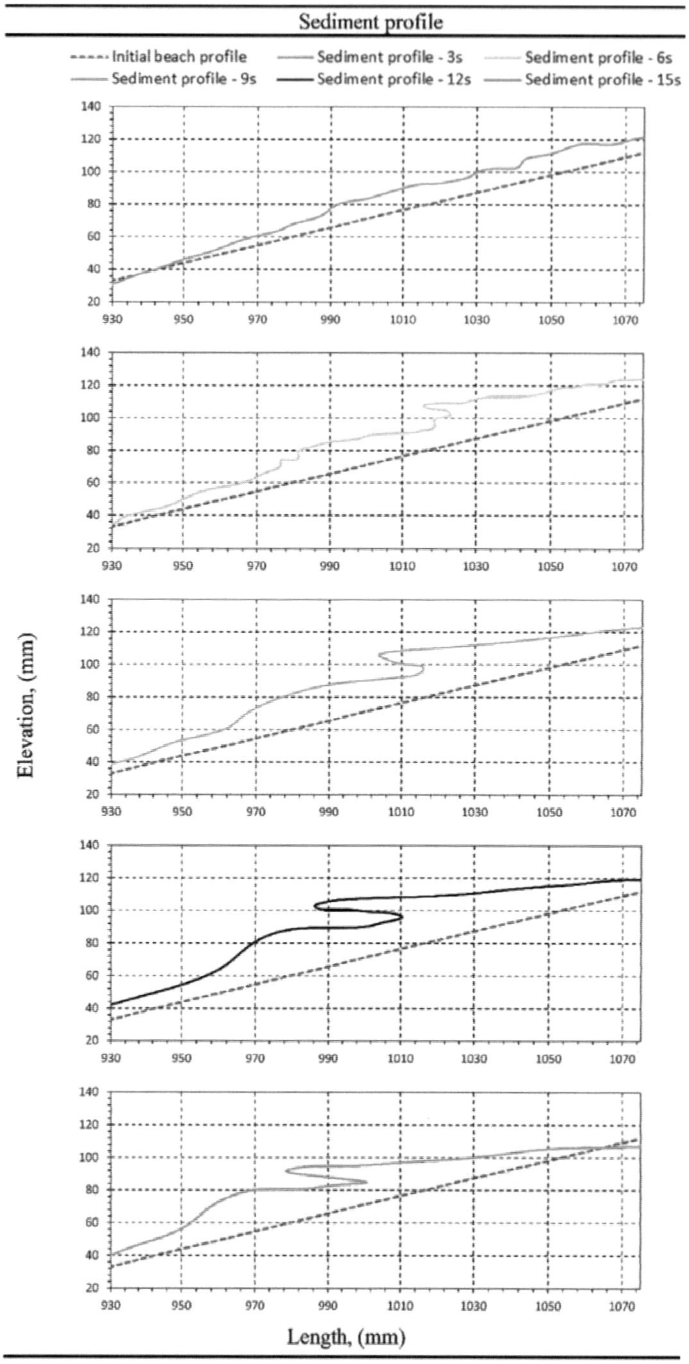

Figure 3. Sediment profile changes within the 15s periods.

5. Conclusions

In this paper, 3D simulation of SPH is combined with the wave generation equation to simulate liquid-sediment interaction. This 3D simulation was shown effectively simulate the effect of sediment erosion due to interaction of wave. Soon, the complex sediment erosion for coastal area can be model to investigate the erosion and accretion.

6. References

[1] Müller M, Charypar D and Gross M 2003 Particle-Based Fluid Simulation for Interactive Applications *Proceedings of the 2003 ACM SIGGRAPH/Eurographics symposium on Computer animation* pp 154–9
[2] Gingold R A and Monaghan J J 1977 Smoothed particle hydrodynamics: theory and application to non-spherical stars *Mon. Not. R. Astron. Soc.* **181** 375–89
[3] Lucy L B 1977 A Numerical Approach to The Testing of The Fission Hypothesis *Astron. J.* **82** 1013–24
[4] Manenti S, Sibilla S, Gallati M, Agate G and Guandalini R 2012 SPH Simulation of Sediment Flushing Induced by a Rapid Water Flow *J. Hydraul. Eng.* **138** 272–84
[5] Shi H, Yu X and Dalrymple R A 2017 Development of a two-phase SPH model for sediment laden flows *Comput. Phys. Commun.* **221** 259–72
[6] Fourtakas G and Rogers B D 2016 Modelling multi-phase liquid-sediment scour and resuspension induced by rapid flows using Smoothed Particle Hydrodynamics (SPH) accelerated with a Graphics Processing Unit (GPU) *Adv. Water Resour.* **92** 186–99
[7] Zubeldia E H, Fourtakas G, Rogers B D and Farias M M 2018 Multi-phase SPH model for simulation of erosion and scouring by means of the shields and Drucker–Prager criteria. *Adv. Water Resour.* **117** 98–114
[8] Ulrich C, Leonardi M and Rung T 2013 Multi-physics SPH simulation of complex marine-engineering hydrodynamic problems *Ocean Eng.* **64** 109–21
[9] Hu X Y and Adams N A 2006 A multi-phase SPH method for macroscopic and mesoscopic flows *J. Comput. Phys.* **213** 844–61
[10] Ghaitanellis A, Violeau D, Leroy A, Joly A and Ferrand M 2015 Application of the unified semi-analytical wall boundary conditions to multi-phase SPH *10th International SPHERIC Workshop, Parma, Italy*
[11] Ghaïtanellis A, Violeau D, Ferrand M, Abderrezzak K E K, Leroy A and Joly A 2018 A SPH elastic-viscoplastic model for granular flows and bed-load transport *Adv. Water Resour.* **111** 156–73
[12] Zainol M R R M A, Kamaruddin M A, Zawawi M H and Wahab K A 2017 Numerical analysis study of sarawak barrage river bed erosion and scouring by using Smooth Particle Hydrodynamic (SPH) *IOP Conference Series: Materials Science and Engineering* vol 267
[13] Violeau D 2012 *Fluid mechanics and the SPH method: Theory and applications* (Oxford University Press)

Acknowledgements

This research was funded by UNITEN R&D Sdn Bhd, Tenaga Nasional Berhad seeding fund under grant No. U-TG-RD-18-05, FRGS grant FRGS/203/PMEKANIK/6071322 and USM RU grant by USM/1001/PMEKANIK/8014071

AUTHOR INDEX